Beginner's Guide to Home Energy Saving

Beginner's Guides are available on the following subjects:

Audio
Central Heating
Colour Television
Computers
Domestic Plumbing
Electric Wiring
Electronics
Home Energy Saving
Integrated Circuits
Radio
Tape Recording
Television
Transistors
Woodworking

Beginner's Guide to Home Energy Saving

Peter Campbell
B.A.Arch., Dip.T.P.

Newnes Technical Books

The Butterworth Group

United Kingdom	**Butterworth & Co (Publishers) Ltd** London: 88 Kingsway, WC2B 6AB
Australia	**Butterworths Pty Ltd** Sydney: 586 Pacific Highway, Chatswood, NSW 2067 Also at Melbourne, Brisbane, Adelaide and Perth
Canada	**Butterworth & Co (Canada) Ltd** Toronto: 2265 Midland Avenue, Scarborough, Ontario, M1P 4S1
New Zealand	**Butterworths of New Zealand Ltd** Wellington: T & W Young Building, 77—85 Customhouse Quay, 1, CPO Box 472
South Africa	**Butterworth & Co (South Africa) (Pty) Ltd** Durban: 152—154 Gale Street
USA	**Butterworth (Publishers) Inc** Boston: 10 Tower Office Park, Woburn, Mass. 01801

First published 1980 by Newnes Technical Books,
a Butterworth imprint

© Butterworth & Co (Publishers) Ltd, 1980

British Library Cataloguing in Publication Data

Campbell, P. M.
  Beginner's guide to home energy saving.
  1. Dwellings — Great Britain — Energy conservation
  I. Title
  644   TJ163.5.D86   79-40101
  ISBN 0-408-00442-8

Typeset by Butterworths Litho Preparation Department
Printed in England by Fakenham Press Ltd, Fakenham, Norfolk

Preface

Progress is measured today in terms of the ability to do more with less — to make limited resources go further. Nearly 30 per cent of Britain's fuel and power is used in houses, many of which are nevertheless heated to a poor standard. Fuel prices continue to rise. However, recent research has proved that new houses, designed and built with a view to economy with energy, can provide very acceptable standards of comfort using far less fuel.

Of the existing stock of houses, many are embarrassingly wasteful of energy. The better the new houses are, the worse the old ones will seem, especially to those who live in them and have to pay the ever-rising bills. Older houses can in fact be improved a good deal, although an improvement across the whole stock will take a great effort, spread over perhaps 20 years. It will have to be done largely on the initiative of house owners, who will need to find out more about heating and other domestic uses of energy.

Research is providing an improved understanding of the complexity of the whole subject and of the interactions between all the factors involved. The resulting knowledge (much of it not widely published hitherto) can now be offered to householders in a form that is both sophisticated and at the same time easy to grasp.

More is known about the way houses behave in relation to heat and heating systems, and there are new ideas about how to save fuel, how to be comfortable at lower air temperatures, and how to exploit the physical characteristics of a particular house, both to warm it more efficiently and also to make better use of natural energy. Some long-standing ideas have come to be revised, about how much energy is used for each of the principal household purposes. The problem of condensation, with its potentially damaging consequences, is also better understood, and more experience has been acquired about insulating materials in practice.

As research work is still going on, some of the advice in this book is inevitably inconclusive; where doubts still exist, they are made clear in the text.

This book was written by permission of the Department of the Environment, as an independent venture based on knowledge gained in the course of work in Government research and development. The views expressed are the author's own, and not necessarily those of the Department or of any other person or body involved in this work.

The author is indebted to the following organisations, contact with whom over the past five years has provided much information.

The Department of the Environment: Housing Development Directorate; Building Research Establishment; Building Regulations Professional Division
The Department of Energy
The British Gas Corporation
The Solid Fuel Advisory Service
The Electricity Council
The School of Architecture, University of Bristol
The University of Newcastle upon Tyne: School of Architecture, Building Science Section
The University of Wales Institute of Science and Technology: Welsh School of Architecture
Heriot Watt University: Department of Building
Milton Keynes New Town Corporation
The National Building Agency
The Consumer Association
The Open University

Special thanks are due to L. J. Pierce of HDD for reading the draft for Chapter 4, and to the Directors of HDD, to W. B. Pascall of RIBA and to Professor P. E. O'Sullivan of UWIST for initial encouragement in the venture. Also to Val for editorial help and to Penny for typing.

P.M.C.

Contents

1 Introduction

For the present, Britain is not short of energy. But the prices we pay reflect world prices, and the world as a whole is running out of the energy resources that we currently rely on. We are looking for new sources of energy, but unless we stumble upon a rather improbable technical miracle all the alternatives will be expensive. Solar energy in its various forms, windpower, wavepower, nuclear power are all expensive in total cost, and some of these are only becoming feasible now that fossil fuels (coal, oil and gas) will no longer be a lot cheaper. In short, fuel prices are more or less certain to go on rising and there will be increasingly good reason to look for ways of using less.

There is guesswork in all predictions, but in Britain the view taken by the Department of Energy in its 1978 Green Paper was that prices will double in real terms by the end of the century. 'In real terms' means in relation to other prices. It would surprise most people to know that British fuel prices have *not* risen in real terms since 1973. General inflation has proceeded as fast as, if not faster than, the average for domestic fuels. We are asked therefore to imagine annual fuel costs at double their present total, out of the same household budget. The prediction of the Green Paper was that they would start to climb steadily from around 1985, the year when world demand was expected to exceed supply.

Britain has begun to prepare for this future in earnest. The Building Regulation minimum standards of insulation for new houses were increased in 1975, and further increases are being considered. Energy-saving Building Regulations are being introduced for other buildings than housing. Energy-saving measures are being encouraged and to some extent subsidised for industry. Money from central government is being spent on most types of public buildings to save fuel. As from

1

1977, money is being made available to local authorities to subsidise the insulation of houses, and since 1978 as grants to private householders for the same purpose.

However, as far as private houses are concerned, Government subsidies are very modest now and are quite likely to remain so for the majority of people. This seems quite reasonable. Energy-saving improvements are at their most efficient if designed specially for a particular house, heating system and family lifestyle. Government-aided improvements must necessarily be simple and relatively crude, and even then are expensive to administer. Where people contrive their own, so long as understanding is brought to bear, more energy is likely to be saved in relation to the money spent. A great deal of time-consuming and irritating paperwork is avoided. Perhaps most important, the responsibililty for what is done rests with the householder, where it belongs.

This book is concerned with energy-saving measures that are 'cost-effective'. It is senseless to spend money that will not be won back in energy saved within a reasonable time, unless it recoups its value in improved comfort or simply in the fun of the job. Following this principle, any improvements you make are in your own interest. The energy saving to the nation is a bonus for which future generations may have reason to be thankful.

Most of these improvements last a long time — many as long as the house itself. Once they are done they will continue to save money, whereas if they are put off they are unlikely to cost less when that time comes, and may well cost more.

If by tailoring your improvements to your own house you can save more fuel for every pound spent, you will be able to do more within the limits of what is cost-effective. The following chapters demonstrate the case for purpose-designed improvements, but briefly the argument runs as follows. A house, its heating system, its pattern of ventilation and the household in it usually add up to a very individual combination, in which all features interact with each other. Rooms are used for different purposes at different times, doors are closed or left open for very good reasons, and heat moves around the house. The sun shines a lot on some houses, while others are shaded by trees or other buildings. Some families smoke and need more ventilation, and so on.

Good insulation, and care about wasting energy, mean that hitherto trivial features of the house's behaviour become more significant. Being comfortable while using less fuel requires understanding, subtlety and ingenuity. It follows that if you try to proceed along the lines of simplified advice given for some non-existent 'average house', there is a large chance that what you do will fail to match your particular circumstances accurately. I am not saying that you will achieve nothing, but you will almost certainly not get the best results for your efforts. There is also the chance — albeit remote — that you will do something harmful.

Your most sensible course of action as a householder is to learn how houses behave, in their use of energy, well enough to apply the principles accurately to your own. Alternatively, you could employ a professional to do this for you; but it would take him or her many days' work at your expense to acquire as good an understanding of your house as you can acquire yourself with a little study and thought and the advantage that you live in it. (This is not to belittle the competence of the professions, but at this time there are probably few who are competent and experienced enough in this field to analyse quickly such complex matters, as accurately as you could yourself.)

It is possible that modern generations know less about the heating of their homes than their grandparents did or do. With laborious and inefficient methods of heating, to be comfortable one had to use a lot of cunning. What has spoiled us is whole-house heating at the touch of a button, and apparently plentiful supplies of fuel, which we have been squandering as the Incas did their gold. However, we have more science at our disposal, which gives us the chance to add this extra body of knowledge to our own commonplace household skills. The prospect of learning it is not daunting: cooking, gardening and house maintenance each contain far more learning and complexity. It is necessary for the future of the industrial world that all ordinary people become 'energy-wise' again.

The 'energy wisdom' that is needed consists of both the theoretical knowledge and practical experience. It involves understanding how much energy is used for each household purpose, even though most of that energy is invisible; how houses lose and gain heat in various circumstances, and how to make rooms comfortable with the least possible expenditure of fuel; how to adapt the house in the course of time to become as efficient as possible, and how to use it in the most efficient way.

Using the fruits of the most up-to-date research at present available, this book shows you how to work things out for yourself.

Chapters 2 and 3 go into the facts you need to understand before doing anything at all. If as a result you decide it is a waste of effort and money to make any physical changes to your house at all, then you might have done yourself a valuable favour already. But the book also invites you to approach the subject 'strategically'. This could mean:

● doing first those things that produce the best fuel savings in relation to their cost;

● looking for opportunities to make changes at the most advantageous time (insulating solid walls when redecorating, draughtstripping windows a week or two after a new paint job, incorporating 'trickle ventilation' in replacement windows, etc);

● observing the principle of matching the heating system to the performance of the house, because this relationship is a vital factor in achieving comfort at low cost.

Chapter 4 deals with heating systems, their use and how they relate to the characteristics of the house. It also discusses hot water systems. Chapter 5 deals with 'routine savings'. This expression refers to the principles of running the house so as to avoid energy waste, at no cost in cash or comfort, just by a better understanding of sunshine, ventilation, the storage of heat and its movement around the house. This advice becomes more meaningful once you have grasped Chapters 2 and 3, and is just as important in the best-insulated houses — perhaps more so.

Chapter 6 discusses lighting and electrical appliances. For many households these two uses of energy account for a large proportion of the total annual cost (Chapter 2 will have told you how to find this out).

Chapters 7 and 8 go into the practical details of the most common improvements, most of which are within the means of a competent handyman. Some less common ideas are included, too, as possible solutions to particular problems. Chapter 9 gives advice on possible technical hazards, to help you avoid making alterations that lead to trouble later on, perhaps many years hence.

Chapter 10 discusses heat from the sun, insofar as this is relevant to today's houses and technology, and offers some useful DIY ideas. Chapter 11 offers what insight is available at present into the more advanced ideas in energy saving that are likely to become common during the next few years. One or two of these might offer experimental ideas for the adventurous, but more by way of a hobby than as a foolproof way of reducing fuel bills. Others (such as heat pumps) involve factory-made equipment, but these are likely to come on the market at relatively low prices before very long.

Hence, parts of Chapter 10 and all of Chapter 11 are for information or for interest, on the grounds that all things concerning energy will have to become common knowlege within the next few years.

Above all, this is a book to be taken slowly. It may be convenient for a few people to plan a complete set of improvements all at one fell swoop. But probably, for most, there is scope for a programme of jobs tackled one by one over perhaps several years, as money and time permit. There is after all quite a lot of thinking to be done, and with thought and experience you will become wiser.

Because the effects of such improvements are cumulative, you will be able to watch your fuel consumption get steadily less as time goes on — though there must be a limit, of course. You may also be making your house more comfortable, which depressingly few are when they are built. This can be a profoundly satisfying process: a comfortable house is something to be proud of, and if you come to sell it, comfort and low fuel bills will contribute to its value, increasingly for the next two or three decades.

2 Some First Principles

The expression 'cost-effectiveness' has already been introduced. It is a bothersome concept in that it provokes a great deal of calculation, the results of which have an aura of accuracy that is quite false. Such calculations include unprovable assumptions (e.g. what contribution does solar gain make to an 'average' house?) and unreliable data (e.g. the thermal transmittance of a particular material in use). The various patterns of people's habits and their heating needs are entirely unpredictable, in any case. Thus there is scope for a great deal of argument, in which no party can prove its case beyond dispute, and over the past few years an inordinate amount of costly time has been wasted in such arguments.

On the other hand, the issues are seldom so imprecise that agreements cannot be reached on *roughly* what the facts are. It has been necessary for the Government to develop official viewpoints about those energy-saving measures that are likely to be cost-effective and those that are not, in order to make policy decisions about large-scale investment. Such viewpoints have to be defensible, but they are never unassailable: that is a fact of life. This sceptical attitude towards facts and figures is reflected in the chapters that follow.

For any individual household, simplified advice about the order in which to make improvements to the house are almost certain to be inaccurate. A far less uncertain way of estimating priorities is to work it out for yourself, from first principles, using your own knowledge of your own house. For example, as a general rule double-glazing, unless done very cheaply, is not cost-effective in Britain. But there are plenty of individual windows or rooms where double-glazing is without question justifiable. Again, some rooms are quite adequately heated most of the time by waste heat from elsewhere that cannot easily be diverted. For such rooms, any insulation would simply cause overheating and would thus be a complete waste of money. In either case the householder is the only person competent to make the decisions, given the theoretical understanding on which to base them.

Where the energy goes

Your understanding of the energy your house uses has to begin with a grasp of how much you use, for what purposes. Recent research has shown that the differences in energy use, between apparently similar households, are astonishingly great. It has also shown that conventional estimates for 'average' households are often misleading. It seems that many modern families use more energy for hot water and for miscellaneous electrical gadgets than 'average' statistics suggest. There are probably three reasons for this. Averages are usually based on statistics for all households, and have recently become influenced by the growing number of one- and two-person households, who tend not to use much energy. Over recent years family households have been acquiring more and more lights, and also large electrical gadgets, many of which use a great deal of electricity: refrigerators and deep-freezes, colour televisions (especially the older types) and automatic washing machines are the most outstanding examples. At the same time modern houses have become more compact and efficient to heat, and the space heating is being supplemented by more 'incidental gains' from the lights and electrical gadgets.

Recent increases in the price of electricity have made it very expensive. So in a modest family house it is not unusual, if gas is used for heating, cooking and hot water, for the annual cost of electricity to exceed that of gas for space heating alone. Most of this electricity contributes incidentally to heating, in the winter. It is a very expensive way to heat, but there is only limited scope for improving that situation.

Study your own bills

The quarterly bills for past years are a useful first indicator of the energy you use. The June-September quarter should contain little or no space-heating costs in most years (more in northern regions, of course). These bills give you a clue to the usually quite steady consumptions for hot water, cooking, lighting and appliances. (Remember to take account of periods when you are away on holiday.)

Cooking usually uses about 60−80 therms of gas or 1000−1500 kWh of electricity a year for a four-person family, and probably varies between households less than any other consumption. Gas appliances with pilot lights consume small quantities of gas constantly, and these can total one therm per week for a central-heating/hot-water boiler and a gas stove, which may be 15−20 per cent of the total weekly gas consumption in summer. Energy used for hot water is likely to be rather more in winter than in summer for the same quantity of hot water used. This is because the cold supply to the system is usually warmer in summer, when the mains supply itself may be warmer, and also because the water has stood for some time in a warm roof space. Insofar as these assumptions are correct, an estimate of the annual fuel consumption for hot water based on the summer meter reading is likely to be an underestimate. However, these factors may be balanced or even exceeded by the reduced efficiency with which hot water is produced in summer, from a gas-fired central-heating/hot-water system (see Chapter 4). Whether your household takes more baths or washes more clothes in summer or winter, only you can know.

Within reason, the more you know about how much fuel you use for what purpose, the better your judgements will be. Frequent meter reading helps a lot, so you will need to learn how to read your meters if you cannot already do this with confidence. Normal meters (clock type) are in fact rather difficult to read. Regular readings are essential to acquiring a grasp of your rates of energy consumption for various purposes: they can be used in the same way as the mileometer and the speedometer of a car (*Figures 2.1* and *2.2*).

From them you can discover quite accurately how much fuel you use for cooking, and how much electricity for lighting

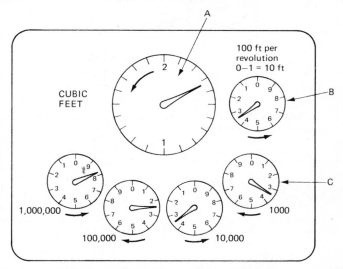

A Large dial (red): needle revolves once per 2 cubic feet, so you can usually see it move, and even time it to learn the rate of consumption. Gas bill gives calorific value of the gas in BTU per cubic foot (usually rather over 1000 for natural gas); 1 therm = 100 000 BTU, so 100 cubic feet approximates to a therm.

B Next dial (red) registers 100 cubic feet per revolution, too small a quantity to be included normally in the reading.

C Other dials (black) record higher quantities per revolution. Some read clockwise and others anti-clockwise; just note the *lower* of the two numbers between which the needle lies (or the number to which it points directly). The reading from this meter would be 823 300.

Fig. 2.1 Face of a typical gas meter

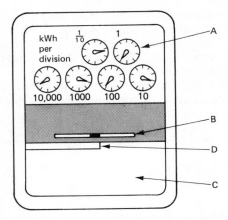

A Dials: the labels 1, 10, 100 etc. indicate the number of kW h per
   division (not per revolution). Half the dials read anti-clockwise,
   the rest clockwise; simply read the lower of the two numbers
   between which the needle lies. The official reader ignores the
   1/10 kW h dial, which is usually coloured red.

B Revolving disc: usually can be seen moving, sometimes quite
   fast. It has a prominent mark at one position on its circumference,
   so you can count its rate of revolution.

C Information plate: includes the vital information (D) as to the
   number of revolutions of the disc that represents one kW h.

Fig. 2.2 Face of a modern electricity meter

and household appliances. Lighting will of course vary seasonally
— in the author's own case, from about 30 kW h per week in
mid-summer to about 70 kW h in mid-winter. Appliances usually
add a lot more. *Table 2.1* is a list of typical consumptions for
common electrical appliances, but of course different makes
and sizes, and circumstances of use, have a bearing on what
happens in practice, so it is of use to monitor your own if you
can.

At the cost of about £7 you can make a simple device for
this purpose, to tell you how much an individual appliance uses
over a day or a week (*Figure 2.3*). You can buy an electricity

Table 2.1 CONSUMPTION OF TYPICAL ELECTRICAL APPLIANCES
(drawn largely from Electricity Authority published figures, with the author's comments added)

Appliance	Indication of consumption	Comments
Cooker	1 person's daily needs per kW h.	Tallies with recorded family consumptions for cookers of 1000–1500 kW h/annum.
Refrigerator	Bench-height simple refrigerator 1 kW h/day.	Upright fridge/freezers 1½–2¼ kW h/day. Freezers 2–3 kW h/day (550–900 & 750–1000 kW h/annum).
Heaters & fires	Usually 1–3 kW, using 1–3 kW h/hour of use.	
Instantaneous water heaters	15 litres (3 gal) of 'piping hot' water per kW h.	'Piping hot' probably means 45–55°C.
Immersion heater	Usually 3 kW.	In practice probably 8–10 kW h per tankful from cold, or up to 6 kW h per large, hot bathful; 3000–4000 kW h/annum for a family is probably realistic, with a well insulated h.w. cylinder.
Television	Black & white 25–150 W continuous load, colour 170–350 W, depending on size and design (some modern sets use markedly less).	Annual consumptions across the range 36–500 kW h/annum, at 4 hours' viewing per day.
Washing machines		Machines vary from 1.4 to 2.4 kW h for the hottest wash and 0.5 to 0.8 for the coolest (all for one 4 kg load). Tumble drying costs extra. Annual costs 200–300 kW h at 2 loads per week.
Lighting		A household awake from 7 a.m. to 11 p.m. experiences about 1620 hours of darkness per year, so each 100 W bulb burning continuously during waking hours uses 162 kW h/annum. Permanently dark or semi-dark areas need a lot more light annually.

meter by mail order from sources advertised nationally. You can use the same device to meter your consumption for lighting separately from other electrical uses, but do not attempt this unless you are knowledgeable and competent with electricity. The consumption you record in one day or week may seem small, but simple multiplication will show its impact over a

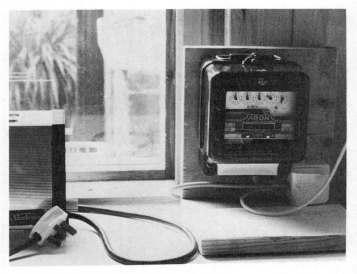

Fig. 2.3 Single-appliance metering device: records the consumption of appliances that draw power intermittently (such as refrigerators) or are used intermittently (such as TVs or lighting circuits)

year. For example, a 100-watt light bulb burned 12 hours a day throughout the year will cost at present £12 a year. (Depending on the meter used, this device may be inaccurate at loads below about 50 W.)

Another, completely free, way of monitoring electricity is to count the speed of the revolving disc on your electricity meter and calculate the kW load being used at that time. The meter indicates how many revolutions of the disc add up to one kW h

Then use the formula:

$$\frac{3600}{\text{No. of seconds for one revolution} \times \text{revs per kW h}} = \text{kW load at this time}$$

(The 'revs per kW h' figure is marked on the meter.) Using this simple technique and switching appliances on and off, you can see how much power is being used by appliances whose consumption you don't have information about. The same technique can be used as a 'speedometer' of the total rate of consumption at certain times.

Knowing how much fuel or power you use for specific purposes may or may not lead you to savings. But it might, for example, lead you to replace some tungsten light fittings with fluorescent ones, or to take special care about the choice of major appliances, or even to relocate some of those you have (especially a deep-freeze) for greater economy. Chapter 6 goes into more detail on this subject.

A gas meter may also be used as a 'speedometer'. Usually there is a dial that moves visibly, if you have a cooker or central heating on, and records 2 cubic feet per revolution. This is not usually read by the Gas Board's meter reader but is used to check the meter. A reference on the gas bill tells you how to translate cubic feet into British Thermal Units (BTUs) and hence therms (100 000 BTUs per therm).

The fact that most meters are hidden in a dark corner and awkward to read makes things difficult for the person who has to pay the bills, which no-one else usually sees. There are psychological problems in educating all the household to take care about energy, which might be more easily overcome if the consumption were readily visible. Some recent experiments in American households have shown that energy is saved if the rates of consumption are clearly visible to the whole household.

It has been suggested that meters should all be prominently placed, with digital readings, and should perhaps include a repeater display that could be reset by a button. They might also include a prominent display of the rate of consumption.

There are American developments in this direction. Many people might regard such a device as a sign of fanaticism; nevertheless children seem to respond to a corporate activity in which they can take part and observe the results. Such education would serve them well: throughout their lives they will have to be far more energy conscious than their parents have had to be.

The first reaction to understanding how much you spend, on what, may be rather discouraging. Many people will realise that space heating does not cost as much as they thought, and the potential for saving from any particular course of action is often relatively small. Subsequent chapters will show how, if you reduce heat loss, the saving will be realised out of the sum of space heating fuel and 'incidental heat gains' which combine to heat the house. Hence, if a planned saving is completely successful, its impact on space-heating fuel consumption may well be greater than expected from a calculation based on that consumption alone. Nevertheless, for most households the only way to make a sizeable reduction in annual costs is by a proliferation of small savings, some by alterations to the house and some by changes in routine. Very few single investments have a really large impact on their own.

Gas and electricity tariffs

The various tariffs are designed to represent reasonably fair means of covering the suppliers' costs. These include the cost of fuel and also overheads — pipe systems, cables and transformers, administration and staff, and the replacement of major sources of supply such as power stations and gas rigs. Because a large proportion of these costs are the same irrespective of how much electricity or gas is sold, it frequently costs you less per unit (kW h or therm) if you use more. This is irritating if you are trying to save fuel, since your savings will normally be realised at the cheapest rate for the fuel. This is because the largest savings are to be made in space heating, and the larger part of space-heating fuel is bought at the cheapest tariff. You are not likely to reduce consumption enough to bite into the few units of fuel you buy at the more expensive tariff. If, however, you can arrange to make economies that actually reduce consumption at the more expensive rate, those economies are worth that much more. For example, while savings in gas are usually made at the normal (central heating) rate of 15.3p/therm (valid at time of writing), if you can in any one quarter reduce your consumption to less than the number of therms for which you pay the higher rate (about 22.8p/therm) you will save at that rate.

With electricity, a far cheaper 'unit rate' applies during the 'off-peak' period for those who can manage to use a lot of electricity at night. However, if you achieve savings from power you draw from the meter during the on-peak (daytime) period, those savings will be worth 2½−3 times as much.

Off-peak tariffs and meter arrangements have changed quite frequently over the last few years. Make sure you have the most advantageous arrangement. Check whether your meter arrangement allows you to pay the lower rate for *all* electricity used at night, and if not, whether a change can be made. Make sure first that a change would be worthwhile all round.

Tariff differentials

Even at the cheapest rate, electricity costs more than gas at present, per unit of useful heat. The cheapest electricity costs about 1.0p per unit (kilowatt-hour) or £2.78 per gigajoule. At the domestic tariff of 2.9p/unit it costs £8.06 per gigajoule (see Appendix 2). The cheapest gas costs 15.3p per therm, or £1.45 per gigajoule, but can only be burned at 50−70 per cent efficiency (other than for cooking). Hence in practice gas costs £2.00−£2.90 per 'useful gigajoule'. Electricity is normally regarded as 100 per cent efficient, i.e. you get 1 kWh of heat per kWh of electricity. In practice off-peak storage heating is not so efficient, in that you usually get some of the heat when you don't need it.

There is no certainty that the present difference between gas and electricity prices will remain. Most predictions of the likely trends suggest that gas prices are likely to increase faster than

those of electricity. Electricity is so expensive now because the
present power stations are only 25—30 per cent efficient in
the conversion of fossil fuels to electricity. (In the UK, hydro-
electric and nuclear stations are only a small proportion of the
whole.) Electricity generation will tend to become a little
cheaper as newer and more efficient power stations replace the
older ones. There is hope also that the latest developments in
nuclear engineering may produce noticeably cheaper electricity.
Gas, on the other hand, will tend to become more expensive as
sources of natural gas are used up, and plant to make 'synthetic
natural gas' from coal or oil has to be built. All such changes are
slow, as major energy-supply installations take several years to
plan and build.

Typical annual costs

The Department of Energy published in 1978 and 1979 an
excellent free booklet, *Compare Your Home Heating Costs*, in
which typical running costs are given for various sizes of house
and types of heating system. As the booklet points out, however, there is a great deal of variety in practice, depending on
household routine, lifestyle and standard of living. Recent
studies of houses in use have shown that gas consumption can
vary from 200 or 300 therms a year for space heating and
cooking in a small house, to 2000 or 3000 therms for heating,
cooking and hot water in a large, well-heated house. The more
you use, the more scope there usually is for saving.

For electricity, official statistics indicate that an average of
1600 kW h is used annually for lighting and appliances, excluding
cooking. This however is an average for all households, including
a large number of one- and two-person households. Government
experience in monitoring family council houses was that annual
consumptions of 2500 kW h were about average for those who
did not cook by electricity, and 3500 kW h for those who did.
These were young families with relatively few appliances, but
a majority were at home all day with young children. Old,
high-consumption colour television sets probably helped keep
these figures high.

The author's own annual consumption is usually around 4000 kW h, but this includes a little electric radiant heating for 'topping-up' in living rooms in cold weather. Also the house is quite large, with a dark internal staircase that needs lighting much of the time, and much of the other lighting is rather inefficient at present. Again, the household's pattern of life is such that most evenings its members disperse to different rooms — a practice that inevitably uses a lot of lighting.

Thermometers

Besides examining in detail your fuel consumption for various purposes, it is a good idea to become acquainted with the temperatures you habitually keep in each room. Use a lot of thermometers and read them often. Put one in every room as a permanent fixture, and one outside somewhere easily visible. The latter must be out of the sun and preferably on a wall not liable to be warmed by conducted or radiant heat from inside — for instance, not next to a window.

Alcohol thermometers are the easiest to read, and can be bought cheaply from chemists. Little square plastic ones of the needle type, made for the dashboards of cars, are even cheaper. They seem to be quite accurate but are not easy to read except from close to and in a really good light, which is not always compatible with the right positioning. Most alcohol thermometers are calibrated in both $^\circ$C and $^\circ$F, which helps one to get used to the $^\circ$C notation invariably used in energy calculations.

The choice of location for a thermometer is critical, as you will discover if you place a number of them around a room. You are likely to find quite large variations according to where the thermometer is. There is probably no perfect place (other than hanging in the middle of the room at chest height), so choose an inside wall out of direct sunlight and well away from any source of heat, including lights and electronic equipment. Thermometers need to be noticeable as you move around the House, or the main object of learning how the house behaves will not be achieved.

Do not fix them all permanently, because you will need at least three to study temperature gradients in rooms, in pursuit of efficient comfort. See Chapter 5. For this purpose they can easily be fixed temporarily to a wall or a piece of furniture with blobs of removable plastic adhesive, without leaving marks afterwards.

Hygrometers

It is worth mentioning this other invaluable household instrument at this stage (*Figure 2.4*). It measures 'relative humidity', which in a house is usually between 40 and 70 per cent. The

Fig. 2.4 Room thermometer (left) is quite handsome and commendably conspicuous and easy to read. The hygrometer (right) indicates relative humidity; cheap instruments of this sort are not very accurate and cannot be used to predict condensation

higher it is in cold weather the more you should worry about the possibility of condensation. A few hours of high humidity are of no consequence, but if it stays high, particularly in a warm house, you should start to worry.

Table 2.2 shows the relationship between air temperature, relative humidity (RH) and 'dew point' (see Appendix 2). It thus indicates the conditions in which condensation will occur. For example, in a room at 19°C, there will be condensation on surfaces at or below 5°C if the RH is 40 per cent, but if it is as high as 70 per cent, condensation will form on surfaces as

Table 2.2 DEW POINT (°C) VARYING WITH AIR TEMPERATURE AND RELATIVE HUMIDITY

	Relative humidity (%)						
	40	45	50	55	60	65	70
Air temperature (°C) 6						0	1
7					0	1	2
8				0	1	2	3
9			0	1	2	3	4
10			0	1	2	4	5
11		0	1	2	3	5	6
12		0	1	3	4	6	7
13	0	1	3	4	5	7	8
14	1	2	4	5	6	7	8
15	2	3	4	6	7	8	10
16	2	4	6	7	8	9	11
17	3	5	6	8	9	10	11
18	4	6	7	9	10	11	12
19	5	7	8	10	11	12	13
20	6	8	9	11	12	13	14
21	7	8	10	12	13	14	15

warm as 13°C. As air with a particular water content is warmed, its RH drops. Hence the colder the air is outside, the lower its RH will be when it is warmed up to room temperature inside. The milder it is outside, the more relatively humid inside; however, in these conditions the surface temperatures of the critical places (walls in stagnant, cold places) will not be so low. Remember, though, that in milder weather these walls will be fewer degrees above outside temperature than in colder weather. You should understand all these factors better after reading Chapters 3, 7, and 9.

Unless you are a scientist with a mind highly tuned to the three factors involved in condensation risk, hygrometers serve two purposes only:

● to warn you when humidity is becoming high enough to warrant your worrying about condensation (i.e. RH above about 60 per cent in cold weather), and

● to warn you when the house seems to be responding too quickly to an increase in moisture (e.g. when cooking or drying clothes), which is a sign that the available absorbent surfaces are near saturation and no longer help modify humidity.

Both conditions need more ventilation or less moisture. The second needs steady ventilation over a long period to correct it. Both are worth being warned about, sufficient to justify having a hygrometer or two in the house to keep an eye on.

Planning for energy saving

Having become acquainted with your patterns of consumption and what various uses of energy cost you, you will then be able to consider where to try to make savings. Give this a lot of thought, along the following lines.

First, see what savings you might make at no cost at all, simply by better management and better understanding of the house, the heating system and ventilation, and the hot water system and electrical appliances. Unless there is some low-priced physical improvement (such as lagging the hot water tank) that obviously should be done immediately, it is not a bad idea to postpone spending money until you have seen how much you can reduce consumption without cost. Inasmuch as you can make savings that way, the potential for saving by methods that cost money will be less. You may decide that this sort of economy is all you want to make. If reading this book has helped you to that decision, it will have served its purpose.

If, however, you decide to go ahead with physical improvements, the next step is to work out which possibilities seem likely to give most return on investment, and do not cost much. Accept the likelihood that your house and living pattern are probably unique, and don't go by generalised advice about what improvements to make first, based on someone's idea of an average house and household. The chance that such advice will match your own circumstances accurately is really rather small, and you could easily waste time, effort and money. In particular, treat all commercial advertising with suitable scepticism.

Advertisements for energy-saving devices or materials rather naturally assume high figures as estimates for your consumption — usually for space heating. When you work out what you really spend on heating, even after including for 'incidental gains' that do not come from heating fuel as such, you are quite likely to find that the advertiser's claims are exaggerated. This is particularly so if you are not able to control the temperatures of all rooms so as to ensure you receive none of the benefits of insulation in increased warmth rather than reduced fuel consumption.

Having decided on the simple and cheap improvements to make first, you may then envisage some more expensive ones to follow when time and money permit. You may feel the need to monitor consumption and temperatures in more detail before deciding about a sizeable investment.

In addition, some improvements may be particularly messy or disruptive, and would most sensibly be done simultaneously with something else that needs doing. For example, the time to insulate solid walls in any room is likely to be when the room is due for complete redecoration. The draughtstripping of windows and doors might well be postponed until shortly after they are painted (allowing about a week for all the solvents to dry out of the new paint). The renewal of part of a heating system gives an opportunity for a major energy-saving change that should on no account be missed. It might be sensible purposely to under-provide with heating either the whole house or part of it, and then 'insulate down' so as to reduce the heat losses to what the new system can match. In this way the cost of the system may be reduced to offset the cost of the insulation, and the new system should be working harder, which (with gas and oil boilers) makes for higher burning efficiency over the season.

3 How the House Behaves

To set about saving energy from heating in your house, you need to understand about the nature of heat and how it behaves in relation to the building. To many people this comes easily, probably from school physics. But for those to whom it does not, this chapter should help to avoid misconceptions. It also contains much that has been learned from recent research into houses and their heating and insulation, so should not be skipped.

Heat is a form of energy, not a substance: it tends to disperse itself among substances (and objects), such as building materials and furniture, water, air, etc. In effect, warmer substances and objects always try to share their heat with colder ones, whenever they have the chance to do so. Substances that acquire heat become hotter, which is to say their temperature increases. Some substances require more heat than others to produce the same increase in temperature; for example, a brick takes rather less heat to raise its temperature by a particular amount than does the same weight of water. All these factors play a part in the way houses behave in relation to heat and comfort. Heat disperses by three quite different means, known as conduction, convection and radiation.

Conduction

Consider a lump of solid substance, to one side of which heat is applied. You will quite rightly expect that in due course the other side will start to get warmer, too. This is because the heat is being 'conducted' through the substance, though nothing seems to move. This process happens to the walls of a house when the inside of the house is warmer than the outside.

Substances differ in how readily they conduct heat. Those that don't conduct it easily make relatively better insulating materials, and can be used to help keep the heat in a building. Most of these, however, rely on air trapped in them: air is a very poor conductor of heat, and is at its best if it can be prevented from moving. So the best insulators have layers or fibres to keep the air still (like mineral-fibre quilt, or duck feathers in an eiderdown) or lots of closed bubbles of air (like the many plastic foam materials that are used for insulation). Metals tend to be good conductors, so are used where heat needs to be transmitted quickly between substances that must be kept apart (such as gas flames and water in a boiler, or water and the surrounding room in a 'radiator').

Convection

Most substances expand when they get hotter, so a particular quantity (by weight) then takes up more space and is therefore less 'dense'. Thus warm air floats on cooler air, and rises away from a source of heat. Immediately, that bit of air is replaced by another, which does the same, so a stream of warm air is produced, which can take the heat away and give it up to other, cooler objects. Some forms of heating appliance rely largely on this process to distribute their heat, and are called convectors. But the same process is responsible for the air currents within rooms and houses which constantly redistribute heat. They do not always do this in the most convenient way, so some rooms tend to have a pool of warm air at the ceiling even if the floor is cold. And the heat from warm rooms downstairs tends to go upstairs − which may sometimes be useful, but adds to the problems of keeping the downstairs rooms warm. Windows left open upstairs in a warm house are very wasteful, as convection effects cause the warm air to pour out and be lost.

Air touching a cold surface gets colder and heavier and tends to fall. Hence a large window or a cold wall will produce a down-draught, which can make some parts of the room uncomfortable and may be one cause of a cold floor (though there may be others).

Liquids are subject to convection in the same way. Hence the hottest water collects in the top of a hot water cylinder, from where the hot supply is always drawn.

Radiation

Surfaces of warm objects 'radiate' heat energy in all directions, such that hotter ones give heat to cooler ones facing them. The rate of doing so depends upon surface temperature and colour: dark colours both radiate and absorb heat better than light or shiny ones.

Radiation is quite independent of the temperature of air in between. For example, you can be made comfortable by the heat of a radiant fire even if the air is quite cold. By the same process, heat is constantly redistributed between surfaces in a room, tending to produce even surface temperatures. But any surface (such as a window or an outside wall or a cold floor) from which heat is conducted will tend to stay cold. Surfaces of substances that can store a lot of heat will also stay cold for longer, until eventually they reach the temperature of the rest of the room.

You can actually feel the effect of a large cold surface, because your body radiates away its heat in that direction and receives nothing in return. It is this 'cold radiation' (or 'negative radiation') that is perhaps the most severe effect that large windows have on the comfort of rooms, and is greatly relieved by double (or triple) glazing. But because radiant heat losses are difficult to calculate they do not usually get taken into account in assessing the economic worth of double glazing, much of whose value is therefore often dismissed.

Curtains, too, protect the inside of a room against the negative radiation of cold glass. The outside of the glass loses its heat much more rapidly after dark in clear weather. For this reason it makes good sense to close all curtains at dusk during the heating season. Note, however, that the heat radiated between objects in a room will not pass out through the window, except by virtue of the surface temperature and the conductive properties of the glass. In scientific terms, glass is 'not transparent' to radiant heat from low-temperature sources, although it is transparent to that direct from the sun, just as it is to light. Hence the interior of the room will not lose heat by radiation through a window to a clear night sky, only to the cold surface of the glass.

Occasionally the roof of a house may experience a frost at night in clear weather, even if the air is not below freezing point, just as frost will form on such nights on the roofs of cars, but not on their windows. These are the effects of negative radiation out of doors, when objects lose heat to outer space, the temperature of which is 'absolute zero'. Houses shielded from the sky by trees or other buildings lose less heat this way.

Two general points concern all three forms of heat transfer. The first is that heat is transferred from the warmer thing to the cooler one, until their temperatures become the same. The greater the temperature difference, all other things being equal, the faster is heat transferred. The second is that all three affect comfort; when you get down to fine detail, there is a lot more to being comfortable than just having a suitable reading on a thermometer. Often by working on this fine detail you can arrange to be comfortable at a lower air temperature, and hence more economically.

Where the heat goes

Following these principles of heat transfer, as long as the surroundings of a house are colder than the house, the house will tend to lose heat. Now, an important point is that some of the heat in a house is provided by means other than heating appliances: sunshine, which the house can absorb and store as heat; and the warmth from bodies, cooking, lighting and electrical appliances, and hot water pipes and systems. These are referred to as 'incidental' heat gains. During the summer all this heat is ventilated away; otherwise the house would become intolerably hot. In mid-season, or mild 'winter' weather, there is often enough heat available from 'incidental' sources to heat the house to a comfortable temperature. But the rate at which the house loses heat is proportional to the difference between its inside temperature and that outside; so, as it gets colder outside, more heat is lost.

As soon as the losses exceed the incidental gains, the inside of the house, too, gets colder. If this is uncomfortable you must either add more heat or reduce the losses. The first rule of energy conservation is: *keep the house warm enough, without having to provide 'space heating' as such, for as much time as possible.* In other words, make quite sure you have reduced the losses as much as possible before you introduce more heat. Research has indicated that most people over-ventilate in milder weather during the heating season and, to compensate, use some heating unnecessarily.

The rate at which the house loses heat, for a given temperature difference, also depends a great deal on how well insulated it is. Hence a well-insulated house will be cosy without any heating in colder weather than a less well-insulated one. However, it is worth remembering that if you make any changes that save energy *inside* the house (such as insulating the hot-water tank or acquiring a more efficient refrigerator or TV set) some incidental heat gains will be reduced, and you will need *more* energy for heating (all other things being equal). The arguments for nevertheless making such changes are that you will still make an overall saving, because:
• incidental heat gains tend to be provided from the most expensive fuels (especially on-peak electricity), and
• they tend to take place all round the year, whereas heating is only needed in the colder months.

Providing the 'balance' of heat

To reach the required inside temperature, space heating is needed to provide the 'balance' between the incidental gains and the total heat requirement. In practice most houses will require space heating when the outside temperature is at or below about 13°C (55°F) or thereabouts — the 'balance temperature'. (Houses, and households' requirements, differ considerably of course.) The average winter temperature in the UK is in the area 5–7°C (41–45°F), so the average temperature difference that has to be 'made up' is about 6–8°C (10–14°F).

However, the *rate* of heat loss is determined by the actual difference in temperature between inside and outside, irrespective of where the heat comes from. So the benefit of an insulation improvement should be reckoned on that basis, rather than on the difference between outside temperature and the 'balance temperature'. For this reason it is possible in favourable circumstances to reduce heat losses by, say, 20 per cent and

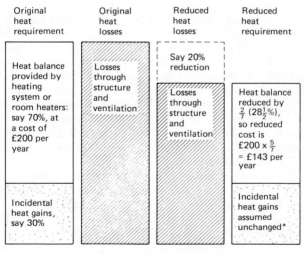

* Incidental heat gains might be reduced as a side-effect of 'internal' conservation measures, or increased by more effective use of solar gains.

Fig. 3.1 Heat balance, and the effect of reducing heat losses (note that the figures mentioned are hypothetical, for illustration only)

be rewarded by a saving on space-heating bills of rather more than that percentage: see *Figure 3.1*.

In extreme conditions the ouside temperature could fall to $-5°C$ ($23°F$) or even lower, and you may need 2½–3 times as much heating power on those occasions. The incidental gains will still make their contribution, though mostly during waking hours, when lighting and 'non-heating' appliances are most in use. In windy weather or with clear skies even more heat may be needed. Do not lose faith in your heating system if on such rare occasions you feel the need for supplementary heating to achieve comfort. In fact most heating systems are sufficiently over-designed to cope, if perhaps at the expense of fuel-burning efficiency in the long run (see Chapter 4).

Rate of heat loss

When discussing, or calculating for, rates of heat loss it is normally sufficient to consider conductive losses through the house structure ('structural' losses) and those through ventilation, rather than to go into conduction, convection and radiation separately. There are conventional methods of calculation that are very simple to use: they enable the merits of insulation methods to be considered theoretically in advance, some indication being obtained as to how much fuel they are likely to save, if all other relevant facts are known. It must be stressed, however, that no great accuracy can be expected from them, for the following reasons:

● They depend on estimates of temperature at certain surfaces, or of the air in rooms, whereas temperatures vary across surfaces and within rooms. Any measurement of outside temperature is similarly approximate, and actual heat losses are influenced by wind speed and negative radiation as well as air temperature.

● They also depend on expressions of 'U-value' (see below) as a measure of the thermal transmittance of part of a building. Even where the U-value of an insulating material can be given accurately, that of existing building materials cannot: for

example, types of bricks and concrete blocks differ, and any one material will conduct heat more readily when it is damp.
● The total of heat loss from the house is strongly influenced by the rate of ventilation, which is very difficult to measure in practice, and impossible to estimate accurately for a particular house.
● Incidental gains are also difficult to measure, as they differ a lot between houses and households.

Nevertheless some sort of calculation method is essential to help make comparative judgements. The conventional methods are accepted as valid for these modest purposes, although the scientists who originated them are well aware of all the limitations.

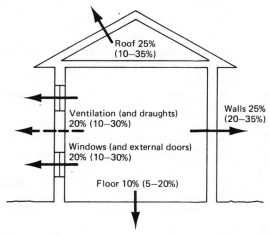

Fig. 3.2 Typical 'percentage of heat loss' diagram (giving common range for each element, in practice); total must always come to 100 per cent

On the credit side, heat loss calculations are a great improvement on the 'per cent heat loss from the average house' formula (*Figure 3.2*), which has been much popularised in the simpler pamphlets and articles on this subject. Some of these also suggest that a proportion of the heat 'stays behind to keep you warm'. That is complete nonsense: the laws of physics make it inevitable that, while the inside of the house is warmer than the outside, heat will be escaping from *all* of it *all* the time. The only issue in question is the *rate* at which it escapes, for a given temperature difference.

Conductive losses are calculated from the 'U-value' of the construction, which is an expression of 'watts per square metre of area per $^\circ$C temperature difference' (abbreviated to W/m^2 $^\circ$C). The watt is a usefully familiar unit, since electrical appliances are rated in watts. This gives you some idea of how much heat you are talking about.

Examples of U-values

A 225 mm (9 in) thick solid brick wall, plastered on the inside, will have a U-value of about 2.0 W/m^2 $^\circ$C. So a square metre of it will conduct away 2 watts of heat for every $^\circ$C of temperature difference. Thus if it is 0°C (32°F) outside and 20°C (68°F) inside, the heat loss of that square metre of wall will be 2×20, or 40 watts. A room with one such wall 2½ metres high and 5 metres long (about $8'4'' \times 16'8''$) will conduct away:

$$\underset{\text{(U-value)}}{2} \quad \times \quad \underset{(^\circ\text{C temp. difference})}{20} \quad \times \quad \underset{\text{(sq. m. area)}}{12.5} \quad = 500 \text{ watts}$$

Such a wall will therefore need ½ kW of continuous heating, to maintain that temperature once it has been reached.

Now consider a whole room (*Figure 3.3*). For this calculation we shall need some other U-values to work from.

Single glazing   U = 4.5 W/m^2 $^\circ$C (ignoring frame)
Suspended timber floor   U = 2.2 W/m^2 $^\circ$C (ignoring joists)
   (20 mm softwood boards)
(Assumed) ceiling   U = 1.5 W/m^2 $^\circ$C

The underfloor temperature is guessed at 12.5°C (55°F) for this outside temperature. It will probably vary a great deal between houses, the ventilation rate (to outside) of the underfloor space

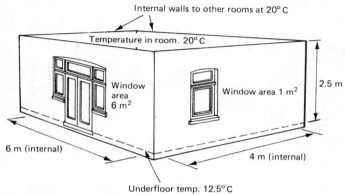

Temp. of bedroom over: 15°C

Internal walls to other rooms at 20°C

Temperature in room. 20°C

Window area 6 m²

Window area 1 m²

2.5 m

6 m (internal)

4 m (internal)

Underfloor temp. 12.5°C

External air temperature: 0°C

Fig. 3.3 Room chosen for calculation example

being crucial (but inadequate ventilation there is hazardous to the structure).

Ventilation rate is calculated approximately by taking the cubic volume of the room (or house) and multiplying by 0.33 × the number of air changes per hour (typically from ½ up to 3 according to relative draughtiness; ½ would be a bit stuffy).

Do not regard this room as necessarily typical, but from these calculations this room ought to need about 2½ kW of heating (from appliances + incidental gains) to maintain a temperature of 20°C (68°F) with an outside temperature of 0°C (32°F). The use of curtains at night would reduce this somewhat, depending on their thickness and closeness of fit.

Considering improvements

Having done those calculations for a room, the next step is to consider what proportion of the total heating requirement is attributable to each element of the room's heat losses. The percentages work out as shown in *Table 3.1*.

Table 3.1 HEAT LOSSES FROM ROOM SHOWN IN FIGURE 3.3

Element	Area (m²)	U-value (W/m² °C)	Temp diff. (°C)	Heat loss (W)	Heat loss (%)
Wall (excluding windows)	18	2.0	20	720	30
Floor (suspended)	24	2.2*	7.5*	396	19
Ceiling	24	1.5	5	180	8
Windows (single-glazed)	7	4.8	20	672	28
Ventilation (1 a.c./hour)	60 m³ × 1 a.c./h × 0.33		20	396	17
Total				2364	100

* U-value of 2.2 and underfloor temperature 7.5°C below room temperature are equivalent to conventional U-value of 0.83 and external temperature of 0°C.

Note that insofar as the estimates are accurate these proportions are valid whatever the average *temperature* in practice. The value of any proposed improvement will be in proportion to total heat loss, not just to 'space heating' cost (*Figure 3.1*).

Such a list offers a first indication of where the priorities are. But bear in mind that the ventilation rate of a living room in an oldish house could easily be 2–2½ air changes per hour. At the latter rate, heat loss through ventilation would amount to 33 per cent of the (revised) total. Also bear in mind that to consider rooms individually in respect of air change is a bit misleading. If the whole house receives a certain rate of air change, most of the loss of heat will be felt approximately doubly in the rooms on the windy side, where the cold air coming in would have to be heated up. Those on the sheltered side would not feel anything, because air would flow into them from other warm spaces and then out through their own leakage to the outside. Bear this in mind when considering airflows, but see also the passage on ventilation (p. 37).

Insulating these external walls with a 'sandwich' insulation board of 25 mm expanded polystyrene and 9 mm plasterboard (a commercial product) would reduce the U-value of the walls to about $0.8 \text{ W/m}^2 \text{ }^\circ\text{C}$. The total heat requirement at this temperature would be reduced by 22 per cent or 432 watts. Double-glazing the windows would reduce the original heat requirement by about 12 per cent, ignoring the question of curtains and any improvement in radiant conditions.

Carpeting the floor would probably have the effect of reducing its overall U-value to about $1.4 \text{ W/m}^2 \text{ }^\circ\text{C}$, and the floor heat loss by 144 watts or 6 per cent of the total. This is perhaps a modest improvement to expect of a carpet (calculated as equal to 10 mm of mineral fibre), but the effect might well be to reduce the underfloor temperature and so partly to eliminate its own improvement.

If you could reduce the air change rate to ½ air change per hour (probably the bare minimum for non-smokers) this would produce 198 watts reduction, or 8 per cent improvement, ignoring the comment on whole-house ventilation above. The sum of these four suggested improvements would amount to a 1054 watts or 45 per cent reduction in heat requirement for the room. Incidental gains, if this were a living room, could easily amount to 500 watts, so the heat 'balance', thus re-assessed as based on 1864 watts, would now be reduced by:

$$\frac{1054}{1864} \times 100 = 57 \text{ per cent}$$

It is interesting to consider that, if you could insulate these walls with 100 mm (4 in) of polystyrene or its equivalent,

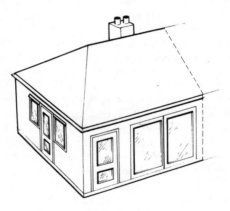

Plan area	8 m × 8 m (64 m²)		
Temperatures	L.R./kit. 18°C, remainder 16°C		
Construction	Cavity walls (U = 1.7); roof (U = 2.0);		
	solid floors (U = 1.0); single glazing		
Heat losses	Walls	1258 W	19%
	Floors	1081 W	16%
	Roof	2162 W	33%
	Glazing and doors	1238 W	19%
	Ventilation (1 a.c./h)	892 W	13%
	Total	6631 W	100%

Fig. 3.4 House A: semi-detached bungalow (2 bedrooms, kitchen and bathroom)

triple-glaze the windows and insulate the floor quite well under-neath, you could probably do without heating altogether, in normal use. But this is not advisable for anyone not well acquainted with building construction, and it would certainly not normally be cost-effective.

Plan area	7.5 m × 8.0 m (120 m² total)		
Temperatures	L.R.s and kit. 18°C, remainder 16°C		
Construction	Cavity walls (U = 1.7); 25 mm roof insulation (U = 1.0); suspended timber floors (U = 0.8); single-glazing		
Heat losses	Walls	3256 W	33%
	Floors	858 W	9%
	Roof	960 W	10%
	Glazing and doors	2960 W	31%
	Ventilation (1 a.c./h)	1677 W	17%
	Total	9711 W	100%

Fig. 3.5 House B: two-storey detached house (4 bedrooms, 2 living rooms, kitchen, 2 w.c.s)

Plan area	6.0 m × 8.0 m (144 m² total)	
Temperatures	L.R.s and kit. 18°C, remainder 16°C	
Construction	Solid walls (U = 2.0); uninsulated roof;	
	suspended timber floors (U = 0.8); single-glazing	
Heat losses	Walls	2455 W	21%
	Floors	682 W	6%
	Roof	2304 W	20%
	Glazing and doors	2968 W	26%
	Ventilation (1½ a.c./h)	3167 W	27%
	Total	11576 W	100%

Fig. 3.6 House C: three-storey mid-terrace town house (5 bedrooms, 2 bathrooms, 2 living rooms, kitchen, etc.)

Figures 3.4, 3.5 and *3.6* show three examples of whole houses, chosen to illustrate how houses differ in terms of the proportions of their total heat loss attributable to the various elements.

Typical U-values

Appendix 5 shows some typical U-values for common construction and common improvements, which should be useful for planning purposes. Note that all such improvements suffer from 'the law of diminishing returns': the more insulation you add, the less return you get for each extra inch (or whatever) of insulation thickness. Really thick insulation is only worth it if either
• you are forced to use a very expensive fuel for heating, and cannot change this, or
• for some reason you heat to an extraordinarily high temperature; or
• as a result you could save quite a lot of money on the cost of a new heating system you are just about to buy, provided you are sure you can adequately control ventilation, too.

Ventilation – why air is needed

Houses need a constant supply of fresh air:
• for people to breathe;
• to allow combustion appliances to burn their fuel completely and supply airflow up their flues;
• to absorb and disperse moisture from domestic processes;
• to disperse unpleasant smells and maintain 'freshness'.
    Having enough air to breathe is never likely to be a problem: in houses of conventional construction, it has been found that more than enough air for breathing alone will always percolate through the structure. Breathing is only threatened when the air is becoming polluted – by smells or cigarette fug, or when a fuel-burning appliance is using up oxygen. In the latter case

the risk is that before the appliance goes out it will burn fuel incompletely and people will be poisoned by carbon monoxide. This is a genuine risk, so treat ventilation seriously.

All combustion appliances need the oxygen in air to burn their fuel. Some modern appliances have a 'balanced flue' straight through a wall, and they exchange fresh air and fumes through this flue; so they take no air from the room and give it no fumes, unless something quite serious has gone wrong. Others (typically solid-fuel stoves) have a separate supply of combustion air through a duct (usually under the floor), which should provide all the air needed. Such appliances again require no air from the room. All others will burn air from the room they are in, and if they have a flue they will need air to keep the flue gases flowing upwards.

Gas appliances are governed by regulations that demand a fairly generous non-closable supply of air to them. Among these are gas cookers, which need have no flue, but again must have oxygen to burn the gas completely. With natural gas the major risk is that the flame will cease to burn and the unburned gas will fill the room and explode, possibly demolishing the house.

The products of properly burned natural gas are regarded as relatively harmless and can be allowed to disperse around the house. However, if they become too concentrated, people prone to lung ailments may suffer. Because the required ventilator can be in any outside wall there is no guarantee that these gases will actually go out through it. Much of the time, air may come in at that ventilator, so the burned gases will disperse within the house.

If this happens to burned gases it will also happen to cooking moisture, smells and dirt. It is for this reason that a kitchen extract fan is desirable. But there must be a fresh-air inlet associated with the cooker as well, to make sure the gas flames are never starved of oxygen.

Moisture, humidity and condensation

Living generates moisture in many ways. People and domestic pets produce it by breathing and perspiring. Cooking, bathing and laundry all produce a lot, especially when clothes are dried indoors, which is the normal practice in houses without special facilities or machinery for it. Room heaters burning paraffin or bottled gas produce a great deal. All this moisture is absorbed into the air as it evaporates. The drier the air to begin with, the more it can absorb. Increasing its temperature increases its capacity further; so even if outside air is very humid, when that air is brought into the house and warmed it can then absorb a little more. All this is familiar to people who have to dry clothes indoors, even if they don't understand the physical laws involved.

When moisture-laden air is chilled it can become overloaded, and will dump some of its moisture immediately in the form of condensation. Condensation on the inside of single-glazed windows is inevitable, in certain conditions, in most houses. How readily this occurs depends on how humid the air is in the house in relation to the surface temperature of the glass.

Usually condensation does no harm, but where it happens continually spots of mould may develop in the wettest areas. Houses that are not well heated, or are badly insulated and not well ventilated, are likely to experience it on walls, floors and ceilings and in cupboards; in any of these places it is ugly and depressing, and possibly also destructive and unhealthy. Dampness can also cause rot — see Chapter 9.

Mould growth

Houses that are kept continuously warm and adequately ventilated do not tend to acquire mould. This is not because less moisture is put into their air but because in periods of high humidity their walls, ceilings and furniture absorb the excess, to release it again later. Thus the humidity of the air never gets excessive.

This capacity to absorb depends greatly on the materials of the building. Gloss-painted walls and vinyl wall coverings absorb nothing. Soft plaster on brickwork is highly absorbent. However, if the house is kept cold the air itself will always bear less moisture, so the chance of a 'safe' rhythm of humid and less humid periods is reduced. If absorbed moisture does not have the chance to dry out between humid spells, then the house's capacity to absorb will diminish until one day condensation will start.

Mould spores exist in the air all the time, and they only require surface wetness or dirt to germinate. After that, mould will grow whenever the relative humidity exceeds about 70 per cent (see Chapter 2), which occurs quite frequently in most houses. Unfortunately mould is inevitable in some houses because the occupants simply cannot afford to keep them both warm enough and well enough ventilated to avoid the conditions that start it. Old houses with high 'thermal capacity' (see below) may never have had mould before, having been heated by solid fuel which ensured a constant background warmth and continuous ventilation. Modernisation with central heating systems that can be run intermittently in an attempt to save fuel can cause the behaviour of such houses to change.

Another cause of mould is replastering with harder plaster, which reduces the walls' absorbency. Furthermore, when building work is done a great deal of water from 'wet' building processes will be left in parts of the structure, sufficient to get mould started. The same applies to new houses, which may take a year or more to dry out properly.

So long as the house is kept warm, not too much ventilation is needed. The best sort is a small amount of 'trickle' ventilation (see Chapter 7) continuously, to be closed only when it is so windy that structural percolation alone is sufficient. Opening windows nearly always provides far more ventilation than is needed in the winter, and if you rely on windows you are likely to want to close them when you are out, for security reasons. Trickle ventilation has a minimal effect on heat loss, and provides a steady means of removing moisture stored in the house from the last period of high indoor humidity. It also disposes of fug and smells at a slow, steady rate.

The mechanics of natural ventilation

For air to move into and out of a house, or through it, there have to be holes and there has to be air pressure. The simplest process to grasp is that of a single house in a wind, with raised pressures on one side and reduced ones at the other, so that air gets pushed and pulled through the house in the same direction as the wind.

In practice the process is seldom so simple, because even in wind quite complicated patterns of higher and lower pressures develop around the house, depending on its shape, the wind direction, and any other nearby objects that create turbulence. A house may develop habitual patterns of airflow, which you can discover by observing (with a small source of smoke) where draughts come in and where they go out. But these will not be the same in all conditions. Sometimes when there is very little wind some draught gaps will puff and suck alternately, having a small effect in providing fresh air, but causing no noticeable cold currents in the house.

More often than not there is a tendency for suction currents to predominate at the roof, so a loft space will have air dragged

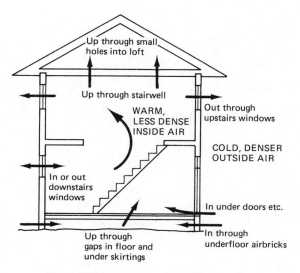

Fig. 3.7 Buoyant air movement ('stack effect')

out of it much of the time, if there are ventilation holes or gaps. This will tend in turn to pull air out of the house through holes in ceilings (and perhaps up from hollow partitions), and can lead to a continuous and very wasteful heat loss, not noticeable in the rooms from which the heat is lost. The buoyancy of warm air increases this tendency in the winter.

As warm air rises, it produces a 'chimney' or 'stack' effect in a space, increasing in proportion to the height of the 'stack'. Hence air at the top of the stack is under pressure to escape, pushed by the equivalent columns of cold air outside, which is heavier, trying to get in lower down (*Figure 3.7*).

It is important to be aware of this effect, because it is a major cause of heat loss: warm air leaks from high windows and from bedroom and bathroom windows, even when there is no wind. As long as there are ways out at high level there will tend to be

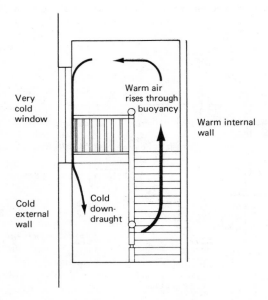

Fig. 3.8 Common air movement patterns in a stairwell

continuous cold draughts at low level, for example between floorboards and under skirtings downstairs.

The same process is responsible for much of the supply of warm air to the bedrooms of houses heated on the ground floor only. It can also cause constant up-and-down airstreams in stairwells that have a large window or cold wall to cool the air near the top and a source of heat near the bottom (*Figure 3.8*).

Heat losses from ventilation

The better a house is insulated, the greater will be the proportion of the total heat lost through ventilation. The loss itself will not be greater, but discomfort effects from cold draughts could be more severe, especially where heating systems are suitably scaled down.

The calculations for ventilation heat loss (*Table 3.1*) were based on the amount of heat it takes to raise the temperature of 1 cubic metre of air by 1°C. (In fact, moister air takes rather more heat, and drier air less heat, but the figures chosen are adequate for these purposes.) It is conceivable for ventilation heat losses to amount to a quarter or even half the total losses from a well insulated house, so either ventilation has to be much better controlled or heating systems must be made much more flexible and responsive to cope with rapid changes in air leakage. The techniques available for reducing ventilation rate are discussed in more detail in Chapter 7. They include:

● draught sealing of all windows and doors;

● identification and cure of all other gross leakage paths, which are likely to amount to a good deal more than those around windows and doors; particular attention should be paid to leakage paths between the house and the loft space;

● provision of controllable trickle ventilators to ensure small and steady airflows that can be adjusted to outside windspeed;

● fitting of draught lobbies to all outside doors to prevent the shock effects of opening these doors in cold weather;

● the possible introduction of staircase enclosures or curtains to reduce warm airflow to the upper floor or floors.

Heat transfer to upper floors

Where roof insulation is added to a house heated on the ground floor only, bedroom temperatures will inevitably rise, and as a result *more* heat will be lost through bedroom walls and windows. The result is usually a disappointing return on investment in roof insulation, but warmer bedrooms, which you may or may not welcome. Conditions more comparable with whole-house heating can thus be achieved from a ground-floor-only heating system.

This rise in bedroom temperatures is even greater if their walls also are insulated. The fuel saving will then be much greater, but still less than you might have expected if you did not take the temperature changes into account. See *Figure 3.9.* (Note that all the figures used here are for illustration only, and are not likely to be valid for any particular house. The rate of internal heat transfer depends on the design of the house, and also to a very great extent on how you run it.)

Although roof and wall insulation will produce warmer bedrooms, the degree of control you have over their temperatures will be minimal, or at least crude. Some of the heat they receive will come up through the floor from below, but most, by 'stack effect', will come up the staircase. You can control the latter to some extent by the opening or closing of doors, and if possible by closing off the bottom of the staircase with a door or curtain. But you should realise that wall and roof insulation can affect internal temperatures considerably wherever they cannot be controlled.

Insulation may, for example, make redundant some or all of the upstairs heating in a 'wholly heated' house. At the very least it will require adjustment of upstairs heat emitters to ensure that the value of the insulation is not being wasted in higher temperatures. Adjusting household routines to the new situation may be irksome; but make sure such adjustments are not achieved by simply opening bedroom windows more.

There are disadvantages to having bedrooms heated purely by conduction and convection from downstairs. The convected part of the heat supply to upstairs will tend to increase when

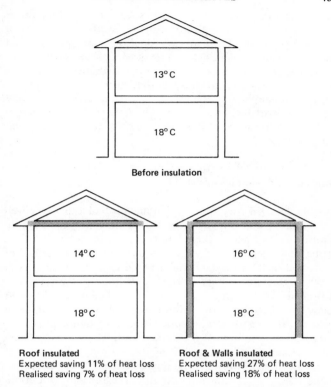

Before insulation

Roof insulated
Expected saving 11% of heat loss
Realised saving 7% of heat loss

Roof & Walls insulated
Expected saving 27% of heat loss
Realised saving 18% of heat loss

Fig. 3.9 Effect of warmer bedrooms on savings resulting from insulation (no upstairs temperature controls)

colder weather tries to make bedrooms colder, because the driving force is 'stack effect', which becomes stronger in those conditions. But the more internal air movements of this kind there are, the more cold draughts are likely to be created within the house. Whether they cause discomfort or not will depend on the design of the house. For some, the only options open may be to fill the whole house with a generous supply of warm air or curtain off the staircase and endure cold bedrooms. The former will be far easier in a well sealed and insulated house.

It is in everyone's interest that such crude techniques as using doors to control upstairs temperatures become a normal part of household life. If house insulation improves, the total heat requirement might become small enough to dispense with the rather complex and expensive heat distribution systems we use now, but simpler systems will only be acceptable if people realise that simpler, 'people-dependent' heat distribution techniques are workable without loss of comfort. Such techniques have the crowning advantage that they contain nothing that can go wrong.

Comfort, and what it consists of

Anyone who has used room thermometers will know very well that comfort depends on other things than just the air temperature at the place where the thermometer happens to be. To be comfortable, thermally, we need a suitable balance of convected heat (warm air) and radiated heat, the meanings of which were explained earlier in this chapter. We also need air movement within upper and lower limits. Too little movement makes us feel stale and stuffy; too much makes us feel cold even if the air is at a perfectly acceptable-sounding temperature.

Professor O'Sullivan of the University of Wales, in a television talk, defined comfort as 'lack of discomfort'. The secret of energy economy is to have everyone in the room comfortable at the least possible consumption of purchased energy. Certain rooms in the house will of course be crucial in this respect, namely those in which people sit still for long periods, and where they are thus most sensitive to thermal comfort. An important element of the process of making useful economies is that of studying these rooms in some detail and discovering how to make subtle improvements that will allow them to be comfortable at a lower temperature. These again are problems that only individual households can solve, although some hints are offered in Chapter 5.

Simple monitoring will be needed — perhaps by using several thermometers (*Figure 3.10*), by using candle flames or cigarette

Fig. 3.10 Three thermometers set up to identify temperature gradients in a room. In this room heated by a large radiator with no shelf, the thermometers read (from the floor up) 17°, 20°, 23° C; there is an unpleasant cross-draught at floor level, making for cold feet

smoke to indicate predominant air flows, and by feeling surfaces to check for cold ones that will cause uncomfortable negative radiation, requiring warmer air or a radiant fire to counteract it. Quite minor improvements at nil or trivial cost may save you a degree or so of air temperature, which over a whole winter is worth a great deal.

Heat storage and thermal mass

The final important principle to grasp, in understanding the way a house behaves when heated, is to do with its capacity to store heat. The relevant factors are the storage capacity of the materials it is made of, and their conductivity. There is so much variety in the subtleties of behaviour that in practice it is only of use to describe houses in terms of their relative speed of response to heat that is introduced. The simplest expressions are 'high' and 'low' thermal mass, or 'heavyweight' and 'lightweight' houses – all such terms being relative.

The effects of these characteristics in practice are felt in terms of how quickly the house will warm up 'from cold', and how long it takes to cool down from comfort temperatures after the heating is switched off. As well as the thermal mass of the house, critical factors are:
● the thermal mass of furniture in it;
● the power of the heating system;
● the rates of heat loss (through the structure and through ventilation) at the time, depending on insulation and how cold it is outside.

Clearly, a 'heavy' house will take a long time to heat up from cold, especially in cold weather. If you wish to heat such a house intermittently you may find it takes so long that in cold weather you can only afford – without being uncomfortable – to switch the heating off for quite short periods. But in milder weather you may be able to warm it up to a comfortable temperature for the evening, and having switched off at bedtime find there is still enough heat left next morning to need no extra boost to have it warm enough for getting up, as long as everyone is going to be out all day.

A 'lightweight' house, on the other hand, may cool so quickly that you need to add heat at some time during the night to avoid waking up feeling cold early in the morning.

If you go away for a spell in the winter, leaving the heating off, it may take even two or three days to heat up the walls of a heavy house enough for the living rooms to feel comfortable without additional radiant heat. The effect of cold walls is most keenly felt in terms of these radiant temperatures, even if the air temperatures seem to be high enough for comfort.

Insulating the walls of a house (whether on the inside, or in the cavity, or externally) will have a profound effect on their performance. Generally, improved insulation will 'slow down' the response of the house in most conditions. Of course a 'heavy' wall will heat up rather more quickly if it has insulation on the outside than if it has not, since while a lot of heat will still be needed to warm it up, less will be leaking out. Such a wall with external insulation will then cool down very slowly indeed — a useful characteristic if you have night storage heaters, making it easier to retain the stored heat all day.

The storage capacity of the house structure can be exploited to help run a gas-fired radiator system more efficiently. You can cause the boiler to run at full blast (and hence at its most efficient — see Chapter 4) until the radiators are very hot, and then switch off, enjoying the heat stored in the structure for many hours afterwards.

Solid ground floors can defeat the desire of a household to heat the house intermittently, unless the heating system is very powerful or has a high radiant content. Hence a house with solid floors and a warm-air system may seem to have incurably cold floors, if these do not have time to warm up between periods of heating. The only ways to comfort in such a house may be:

● to keep the whole house warmer, even when you are out;

● to accept that a radiant supplementary fire in the living room is essential, and acquire one that is economical to run, if possible; or

● to insulate the floor, if carpet alone is not sufficient (see Chapter 8).

Surface insulation

People who have used thin expanded-polystyrene wall insulation frequently swear by it. Its effect is to give a cold wall a relatively high surface temperature and hence add greatly to radiant comfort. In principle this is a good idea; it falls down if the wall behind gets so cold that condensation forms at the back of the polystyrene where humid air condenses on the wall surface. It has relatively little impact on the conductive heat loss of the whole wall, being so thin, but it may well make the room comfortable at a lower air temperature, and hence save energy.

A similar effect has been achieved experimentally with a heat-reflective foil wall covering, faced with a decorative lining of a particular vinyl that is 'transparent' to low-temperature radiant heat. This worked extremely well, but was abandoned as being electrically dangerous — which it would be, if the foil were applied by an idiot, neglecting to keep it clear of switches and power points.

The only way to deal with the factors in your house's behaviour associated with thermal mass is to understand the simple principles involved, and study the speed of the rise and fall of temperature in the relevant places. Remember that air-warmed houses may cool off very quickly at first, then appear to 'level out' at some point at which the stored heat in the structure takes over. After this they may cool very slowly indeed. *Figure 3.11* should help explain some common phenomena. It shows a 'temperature profile' — a graph of temperature rise and fall through time, in this case as drawn by a mechanical recording instrument that reproduces a continuous-line chart on a moving strip of paper. Such traces reveal a lot, in particular the long-term average temperatures (useful for calculation purposes) and the rate of the pick-up and decay of temperature in various circumstances. The profile shown is of a house with a very heavy structure in excessively cold weather. On the coldest days the heating could only be cut for short periods twice daily to maintain a temperature approaching 18°C, and on such days boost from an electric fire was needed to exceed

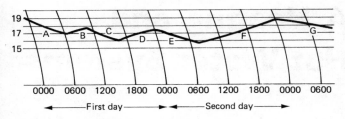

A   Temperature falling after late evening cut-out.

B   Rising in morning after 7 a.m. cut-in.

C   Falling in afternoon after 10 a.m. cut-out.

D   Rising after 3 p.m. cut-in.

E   Falling during second night – very cold outside.

F   Climbing through day (Saturday) as morning cut-out cancelled.

G   Falling slowly overnight as weather turned slightly milder.

Fig. 3.11 Chart record of 2½ days in midwinter: central heating on time clock without wall thermostat. Note how time clock is manipulated to maintain comfort, but boiler is being used at maximum efficiency, exploiting house's heavy structure

18°C. On slightly less cold days the temperature would rise at a steady but slow rate, but would also fall only slowly after the heating was shut down. All slopes are curved: as temperatures rise the rate of rise decreases as heat loss increases; as they fall losses decrease, and also incidental gains contribute an increasing proportion of these losses.

4 Heating and Hot Water Systems

In British houses, heating and hot water systems are frequently combined, and consequently interact with each other. To avoid unnecessary complexity this chapter deals with them separately, (heating first, then hot water systems from page 74), with references to the interactions where appropriate.

Heating: basic principles

To most people, their own heating system is probably so familiar and commonplace that it seems 'normal'; but in fact the range of systems that can be described as commonplace is quite wide, as is the range of their behaviour and control. Because heating arrangements behave so differently in practice — as this chapter will demonstrate — there is good reason to take generalised advice about household energy use with a large pinch of salt.

Recent heating installations have tended to be more limited in their range of hardware, for the sake of lower capital cost and/or greater convenience. Some of these, however, have distinct disadvantages in terms of comfort, economy and reliability.

Whatever the system, naturally there has to be a source, or sources, of heat. 'Central' heating, in its true sense, is a method of heating a series of rooms (and frequently hot water as well) from one source. This is done partly because the source (boiler) is by far the most expensive and bulky item, and concentrates most controls, the flue, maintenance liability and (sometimes) the work of stoking and ash removal, all in one place. This economy is partly offset by either expensive or inefficient heat-distribution methods, usually lacking localised control of room temperature. Distribution is done by water or warm air, which behave quite differently.

The alternative to central heating is to heat rooms separately, each having its own high-temperature source from which heat is distributed within the room by convection and/or radiation in varying proportions. Some room heaters produce enough convected heat to warm the whole house adequately by conduction and convection, as well as the room they are in largely by radiation. This is probably quite a common arrangement in practice, although it is the kind that recent central-heating salesmanship has been trying, with success, to replace. More modern versions of it could well regain popularity in highly insulated houses.

Electric heating is a great deal more adaptable than other forms: underfloor electric heating, storage heaters (perhaps with ducted warm-air distribution), panel room heaters, radiant fires, oil-filled radiators, portable fan heaters and convectors, skirting and ceiling heating and electric blankets are all fairly commonplace. Many can be centrally controlled as well as having room controls. They all behave differently in use, and the list demonstrates the range of possibilities for heating, especially as many of the techniques also have a direct equivalent powered by other fuels. Electricity is the ultimate in convenience and usually is very efficient to use, but unfortunately cannot at present be generated very efficiently; for that reason electric heating is usually expensive to run.

The time factor

Chapter 3 discussed the 'thermal mass' of buildings. This characteristic has its counterpart in a heating system's speed of response to demand, and ideally the two should be considered together. Warm-air systems and gas and electric radiant fires have in common the advantage that they can warm a room — or at least the people in it — quickly. This is a great advantage for people who need to heat their house only intermittently. Such heating will work reasonably well in a 'heavy' house, but rooms thus heated will cool off very quickly afterwards and will be especially susceptible to condensation. They will also feel 'chilly' from their cold walls until the heating has been on long enough to warm them thoroughly — a process that can take several hours or even days.

Very 'heavy' houses will respond more slowly than changes in the heating requirement for any household, and may even be too slow for changes in outside temperature. Thus it may be necessary occasionally to open windows to 'dump' excess heat, or to introduce extra heating to catch up on the increased losses that occur if it suddenly turns cold at night or while the house is empty.

Warm-air systems and radiant fires are at one end of the spectrum of responsiveness; at the other end are solid-fuel systems and fires, which take some time to respond to their controls or to your attention to them. In combination with heavy houses they make for the slowest of all complete arrangements. But people who have lived with them a long time tend to have acquired the habit of thinking ahead and of enduring a chilly house for a short while, and in any case tend not to try to heat intermittently. The two contrasting arrangements described represent extremes, but all combinations of houses and heating have comparable characteristics in some degree, which require to be understood to achieve the best in comfort and efficiency. Whatever sort of system you have, the time factor is important.

How well does the system match your needs? How do you manipulate it to get the best service from it, in particular with regard to fuel economy? The conventional methods of calculation, used to assess and compare systems, tend not to take into account the subtler characteristics of comfort, so it is quite conceivable for a theoretical assessment of performance and running cost to reach a different conclusion from actual experience. So if someone is trying to persuade you to choose, or change to, a particular heating system, don't make up your mind until you have spoken to people with experience of the different systems in question. Get them to back up what they say with figures, and judge how their experience is relevant to your own house, lifestyle and comfort requirements. This may be difficult to do, but try not to hurry the decision.

If a qualified professional gives you advice, bear in mind that his calculations may be some distance from the truth, not because he is incompetent, but because to represent in theory your total circumstances may require a more thorough survey of your house than has been possible; it may even require a computer program to simulate the performance of the system in your house (and even that could be wrong). The chief sources of error are probably the time element and the subtler components of comfort. But always be wary of the person, however competent and sincere, who has an interest in selling a particular kind of heating.

Controllability of systems

'Central' heating systems are by definition centrally controlled, although with some systems a degree of room control is possible. Room control is desirable so that you do not have to keep parts of the house warmer than you need in order to have other parts as warm as you need. Perfect comfort and economy are virtually unobtainable in practice, but in large houses with large heat losses it is cheaper to heat according to need and no more. Smaller, more compact and better insulated houses make such ideals more difficult to achieve, although their overall heat losses may be less. With these, the heat movements between rooms and between floors are more significant.

For example, if you improve the insulation of the house considerably, you may find some rooms become quite warm enough, for most of the time, without individual heat sources of their own. Experiments have shown that highly insulated houses with whole-house heating and thermostatic radiator valves in all rooms have experienced so much heat transfer between floors that bedrooms become warmer, when the living rooms below are well heated, than their radiator valves are set for. At such times — perhaps in practice all the time — those bedroom radiators are redundant.

The above phenomenon apart, room heaters or individual radiators obviously give better control over room temperatures than warm-air or convective systems, simply because warm air by its nature tries to make its way upstairs and heat places that may not be in use. With a warm-air system you cannot readily choose the temperature in any particular room, unless there is a complete duct system throughout the house, with a controllable entry to each room. Many warm-air systems are much cruder, and tend to make unused bedrooms warmer than you would need for most of the time, but not warm enough for prolonged daytime use, as by an invalid or a child doing homework. In such cases supplementary heating would be needed in that room. In practice, however, supplementary heating may be a comparatively economical way of providing additional heating when it is only needed occasionally.

Remember that it is generally a bad practice for rooms to be completely unheated except when in use; such rooms will usually suffer from cold walls, and will need very powerful heaters to make them comfortable. More seriously, they may also suffer from condensation and mould growth, and clothing, bedding and furniture will become cold and damp. However, such conditions are unlikely in a modern house unless it is very intermittently heated, or except in rooms built over unheated spaces.

For most domestic purposes the ideal form of heating is one that provides a constant background warmth in all rooms throughout the winter, with the facility to heat up individual rooms to the required level of comfort quite quickly when they are needed. (In most circumstances 12°C (54°F) should be enough to prevent condensation.) You may be surprised to realise that this description fits quite well a number of heating arrangements commonly regarded as old-fashioned and out of date. Solid-fuel radiant convectors, which were in vogue a decade or two ago, provided just such conditions admirably well, in conjunction with electric fires for occasional 'topping up' of rooms not heated directly. Such stoves allowed considerable control both over their total output and over the balance between radiant and convected heat that they provided. You were better off if your arrangement included two or more of them, especially if one alone could satisfactorily warm the whole house in moderate winter weather. Their main disadvantages were the dirt and physical work involved in loading them and removing ashes, and the problem that the bedrooms of a large house might receive rather unequally the warmth coming up from below. Happily, slow-burning solid-fuel radiant convectors are making a come-back now, and considerable recent research has gone into their design.

Radiant and convected heat, and comfort

The significance of the proportions of its output that a heater provides as radiant and as convected heat has already been mentioned, in the context both of how quickly it achieves acceptable comfort and of how controllable its total output is. Comfort is achieved most quickly where the heat is concentrated on or towards you. So the 'fastest' heat sources are the highly directional ones: the radiant fire with a reflector, or the warm-air blower that can envelop you accurately in a current of warm air. Portable electric appliances do both jobs best, simply because you are free to place and aim them to best effect.

You can be reasonably comfortable with such 'localised' heat sources for long periods, so long as you stay in the right place. The room itself can be quite cold, although in due course it will warm up overall; how soon this happens depends on how quickly it loses the heat you are putting into it. The bigger the room and the worse its insulation, the more economical is such local heating compared with the cost of trying to warm the whole space. Portable electric heaters would thus be ideal for people of very limited means if the cost of on-peak electricity were not so high. (For example, a 2 kW electric fire costs about

5.8p per hour to burn. A gas radiant fire burning at 50 per cent efficiency — about normal — costs 3p per hour at the higher tariff, or 2p at the lower.) An even better low-cost means of survival in cold weather is an electric blanket, of which even less of the available heat escapes from the area of the person needing it. But most of us have come to expect whole rooms or whole houses to be warm, sufficient for the needs of any person in any normal household activity.

Chapter 3 distinguished between air and radiant temperatures, and explained how a combination of warm air and radiant warmth was necessary for comfort. Most forms of heating do in fact provide both, though the balance between the two varies widely. At the extremes of this spectrum are the electric 'infra-red' heater, commonly used in bathrooms, and the warm-air central system. Infra-red heaters can be used to heat large spaces, and have been quite common in workshops and meeting halls, etc., where use is intermittent and there is no need to warm up the structure. They are exceedingly efficient in pure energy terms, in that they provide the kind of radiation that makes you feel warm for the least amount of energy, but the present price relativities between gas and electricity militate against them. In their effect, infra-red heaters can make you feel warm even when the air and your other surroundings are quite cold. In a reasonably warm room they can feel uncomfortably hot.

A warm-air system provides no direct radiation at all, just a stream of warm air. You can treat this as a 'local' heater immediately it comes on, by standing or sitting in front of the output register, but it will quite soon warm up all the air in the room, and thereafter keep it close to a chosen temperature if it is controlled by a thermostat. In Chapter 3 it was explained how the experience of radiant heat is dependent on the size and temperature of surfaces around the body. Smaller, hotter areas can compensate for larger, cooler ones in the same general direction. After quite a long time, depending on thermal mass, a warm-air system will warm up the surfaces of a room and therefore create an improved radiant environment. Until that time you will only be comfortable insofar as the air temperature is higher than you would want it if these surrounding surfaces (walls, floor and ceiling) were not cold.

Heat from radiant sources will also slowly warm the walls, etc. Until it does you will be relying for comfort on the more intense radiation from the source. After the surfaces have warmed up you will want to move away from the source, or reduce its power. Traditional open fires would quite commonly never warm the walls to such an extent that you could not sense their coldness around you. So it was necessary to gather round the fire, sitting in fireside chairs that protected their occupants from the 'cold radiation' of the walls. This was and still is a very workable technique for getting the most from a radiant heat source. A recent invention by Mr Emlyn Richards in Wales — a foil-faced screen for placing around a chair from behind and reflecting the fire's heat back on to its occupant — seems an intelligent advance on traditional thinking.

The 'intermediate' forms of heater provide both radiated and convected heat. Ordinary hot-water 'radiators' actually provide a large proportion of their heat by convection. They can have the disadvantage of causing a strong flow of warm air upwards from them, which in some rooms collects in a pool near the ceiling where the air can be as much as $5-6^{\circ}C (9-11^{\circ}F)$ warmer than that at the floor. The disadvantage of such 'temperature gradients' is that heat is not being provided where it makes for most comfort. Rooms with hot pools of air at the ceiling also pass more heat by conduction to the room above, which may or may not be desirable. Possible remedies are radiator shelves, which help mix the air above radiators, and small fans to push warm air down from ceiling level. As a general rule, small, hot convection sources have the greatest tendency to create hot pools of air at the ceiling. Long convectors of a relatively low power per metre (foot) run produce more even conditions.

A room with hot-water radiators will become comfortable more quickly if the space in front of the radiator is unobstructed.

Sometimes you can feel the heat of a large one from across a room. If you drape clothing over radiators to dry, this will substantially reduce their output, as well as making the air more humid.

Floor and ceiling heating

The Romans introduced floor heating to Britain many centuries ago; theirs was achieved by passing all the smoke and heat from basement furnaces under the floor, probably a very efficient arrangement. More recently floor heating has been achieved by either pipes for hot water or warm electric cables, buried in a solid floor. The latter are cheaper in initial cost and easier to design for even heating, since the cables stay evenly hot along their length, while pipes get cooler since the water in them gives off its heat as it passes through. Underfloor heating is extremely comfortable, especially in rooms not crowded with furniture and without big windows. If electric, however, it suffers all the problems of other electric storage heating (see below). The quantity of heat it can put into a room is limited by a maximum acceptable floor temperature.

Ceiling heating is even cheaper to install than underfloor, but has rather often proved unsatisfactory. In the first place, since it has no capacity to store heat it has to use on-peak electricity, which is the most expensive. Also, some users have complained that the nature of the heat − radiation from above − is inherently uncomfortable. This is certainly so in theory, in that a hot head and cold feet are no-one's idea of comfort. Insofar as the system relies on radiation it will gradually warm the whole room, although the way different parts of it absorb heat will depend on their own nature. A concrete floor with only thin tiles on it will stay cold for a long time, and parts of it in the 'shadow' of furniture will suffer even more. On the other hand, if you can lounge nearly horizontal on a low sofa or chair, you may be blissfully comfortable.

Ceiling heating became popular largely because of its cheapness (in initial cost), which helped to keep down the price of a new house. It is essential for a very thick layer of insulation (preferably about 200 mm or 8 in) to be laid over the ceiling in question, especially in a bungalow with the cold roofspace above.

Storage heaters

Electric night-storage heaters, and underfloor heating metered to perform in the same way, were invented to make use of electricity produced during 'off-peak' periods, when many power stations have to be kept running because they cannot easily be stopped. Those who have such systems can make use of off-peak tariffs during certain hours of the night. The simplest of such heaters have no control over the rate at which heat is given off, so they produce most warmth in the morning and least the following evening, when most is usually needed. A control on them simply governs the amount of heat they take in overnight by controlling their maximum temperature. Thus it is left to your judgement and luck as to whether you guess correctly how much heat you will need for the following day. The new 'Economy Seven' tariff has made for cheaper off-peak unit rates, but in conjunction with shorter off-peak hours. If you change to this tariff, unless your heaters were previously oversized they may prove inadequate in storage capacity during cold spells; in that case you could add to the size of your system, or improve insulation so as to need less heat.

The secret of their effective use is to contain in the house the heat they give out during the day, and this is done by minimising ventilation (the heaters themselves of course require none). A 'heavyweight' house has the advantage that it will absorb heat given out by the heaters and release it later. Good insulation, too, is desirable.

Paraffin and bottled gas

These two types of heater have a lot in common. They are portable and flueless, and carry their own fuel supply. The fuel has to be bought in advance, of course. Both are potentially dangerous if not treated with respect, and both generate a lot of water as they burn, so unless you ventilate the house well you may have problems with condensation. This extra ventilation makes for a waste of heat, of course, so it is difficult to reach fair comparisons with other heaters in running cost. By way of a gross generalisation, however, paraffin is among the cheaper heating fuels, and bottled gas among the most expensive. Both have the advantage that they are proof against most strikes and power cuts, and can be used as a standby in case of the breakdown of a central system. Both types are quite expensive to buy in the first place, compared with (say) portable electric heaters.

Solid fuels

Anyone born before about 1950 was probably brought up on solid-fuel heating of some sort. But then several things happened in quite quick succession to change its popularity. 'Clean air' policies outlawed the open coal fire from most urban areas; electric storage heaters provided a (then) fairly cheap and convenient alternative; and cheap gas central-heating systems, especially the warm-air type, became very popular for new houses. Oil central heating found a market for larger houses, and then the arrival of natural gas made that fuel undoubtedly the best buy, increasingly so ever since. All these alternatives were not only as cheap as solid fuel, but also cleaner and more convenient. Gas and electricity also needed no fuel store and smaller (or no) flues. So people got used to instant, push-button warmth, either throughout the house or at least in part of it.

It has taken some years for the disadvantages of the newer fuels to emerge. The chief of these have been the risk of strikes and power cuts (which affect all electrically controlled and pumped central-heating systems), the equipment's liability to technical failure, and its relatively short life of 10−15 years.

In the meantime solid-fuel appliances have improved from the 25 per cent efficiency of the open fire to around 75 per cent for the latest slow-burning, thermostatically controlled radiant convector/backboiler stoves. Many such stoves can burn ordinary housecoal, which is relatively cheap and plentiful, and consume their own smoke (though not the equally harmful sulphur gases); see *Figure 4.1*. The most modern stoves are well enough insulated to burn all summer in a living room as a source

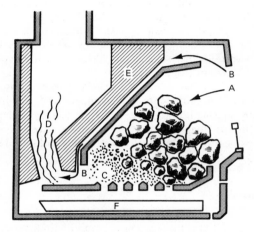

A Air and fuel enter from the top and move down.

B Secondary air passes through channels to the hot base of the fire, C.

D Fed by secondary air, smoke (usually wasted) burns in a combustion chamber at the back.

E A back-boiler forms part of the secondary combustion chamber (the rest being firebrick).

F Removable ash-pan.

Fig. 4.1 Principles of the 'smoke-burning' solid fuel stove

of hot water, without adding so much to the temperature of the room as to justify heating water by electricity instead. At a low output they need attention only once in ten hours, and unlike other common appliances their efficiency tends to increase as burning rate reduces. It is claimed that for central heating and hot water they are cheaper to run than gas appliances.

The same stove will provide for five or six radiators elsewhere in the house by means of a small-bore pump, and the radiant heat that the stove itself provides makes it an excellent living-room heater. It would probably meet all the needs of a modest three-bedroom house insulated to today's standards. Not to be able to 'switch off' the heating entirely might seem a disadvantage to households used to completely intermittent heating, but the constant low background warmth should give a good chance of freedom from condensation.

Provided that the system is carefully designed, the heating and hot-water requirements of a compact and well-insulated house are so small that they invite solving by cheaper and simpler means than those conventionally fitted. This need might well be met by the most modern small solid-fuel stoves.

Woodstoves

In the United States there are now around 800 different wood-stoves available at present, such is the surge of interest created by fossil-fuel shortage. Interest is growing fast in Britain, but is limited by the relative scarcity of suitable timber and by clean-air legislation. There are efficient stoves available from Scandinavia, France, Ireland and now a few from Britain; see *Figure 4.2*. Many of these are exquisite examples of design in cast iron. Good modern woodstoves have 'airtight' construction and accurate air control (some thermostatic), and give the maximum of their useful heat to the room. Some are claimed to 'burn their smoke', but in fact wood is a 'clean' fuel, used correctly in any good stove. In smokeless zones, however, the law requires that any such stove is capable of burning approved smokeless fuel, and some woodstoves are.

The calorific value of timber varies according to its species, but on average is about half that of coal, weight for weight. At the time of writing an average price is about £14 a ton, more or less irrespective of species, as against £50 to £70 for various coals and smokeless fuels. Hardwood logs are best, and a few species are to be avoided. In urban areas, it should still be

Fig. 4.2 Modern French-built woodstove, in enamelled cast iron. Not the most decorative on the market, but it can burn smokeless fuel, so can be installed in a smokeless zone: few other wood-stoves have this facility (courtesy Colin Brownlow & Co, Egham)

possible to salvage 'scrap' timber that has no other use, if you are prepared to cart it and cut it up. There seems to be no general shortage of supply of logs for burning at present, although if woodstoves enjoy a boom in popularity this may change.

It may or may not become economical to grow timber in Britain specifically for burning. Such a development would of course take a long time.

The mechanics of heating controls

This chapter has touched on the question of control in relation to heating systems, but not on controls as such. Controls for space heating serve six fairly distinct functions, all of which can have a very great bearing on fuel economy:

● those primarily for safety, although some of these have a dual purpose in practice;

● those governing the burning rate of an appliance via its air supply (but not usually its fuel supply);

● those controlling the total output of a central system, according to demand ('whole-house' thermostats);

● room controls on room appliances or emitters;

● time controls for pre-determining a pattern of use through a 24-hour cycle;

● zone controls.

Before starting, you should be sure you understand what a thermostat is and does. It controls the temperature of something somewhere by switching heat off (or, with solid fuel only, by damping it down) when a selected upper temperature is reached, and switching it on or up again when the temperature drops to a selected lower level. To make effective use of a thermostat you must grasp both what temperature it 'reads' and what it controls. If it reads the heat distribution medium (e.g. hot water or warm air) it will not effectively control temperature in a room or rooms, no matter what the temperature is outside. Only room thermostats do that. It is a common misconception about the room thermostat that to turn it up will make a room warm up more quickly. It will not do this, though it may help by giving you a higher air temperature than you would normally need, for a spell, to compensate for cold walls. (You would then need to turn it down again later.)

Again, the thermostat will not tell you what temperature you actually have in a particular room. If heat is being added from another source than that controlled by the thermostat, the temperature may be higher than that 'called for' by the thermostat, in which case the 'stat will have already cut out the heating under its control.

Safety controls

Gas and oil appliances have safety devices for cutting out the fuel supply if the flame goes out, for obvious reasons. They are built to be foolproof and one does not hear of them failing. They will be checked in routine maintenance and need never be tampered with. Their only effect on economy is through the size of pilot flames on gas appliances, since these can over the months consume an appreciable proportion of the annual gas bill (see Chapter 2 on monitoring).

Other controls govern the temperature of an appliance to prevent overheating, which could cause damage. The most familiar of this type is probably the boiler thermostat on a gas or oil central-heating boiler, which sets the maximum temperature at which the boiler supplies hot water. Many gas central-heating systems have no other temperature control, because this one in effect governs the temperature of all radiators simultaneously (and, incidentally, of water heated in a cylinder by the same boiler). Because it governs both these services together, without other controls you will find at times that you cannot have both the hot water and the radiators at the temperature you want for each. The simplest way around this problem is to use an electric immersion heater as a 'top-up' facility for the hot water system, for times when you don't need radiators very hot. You will only use this for short periods 'on demand', so it will not cost much. It helps to have the immersion heater switch in a prominent place, to prevent you forgetting it is on. (A neon warning light at the switch or a brightly coloured mark on the switch rocker is also desirable.)

Burning rate controls

Normally gas and oil heating appliances, both of which are fed with fuel under pressure through a pipe, operate either full on or full off (though there are exceptions). The reason for this is that the amount of air going up the chimney matters a lot to economy, but would be difficult and expensive to control in tune with an adjustable flame. Also the heat exchanger (where the flame gives heat to the water) works best for a particular size of flame. The boiler thermostat cuts the boiler out completely when the set temperature is reached, and causes it to light up again later.

Solid-fuel appliances of all types hold a store of fuel in them, and the rate at which it is burned is controlled by the supply of air. Their equivalent to the gas or oil boiler's thermostat is control of the air allowed into the combustion chamber. This makes sure that the boiler does not overheat, at one extreme, or go out, at the other. Ideally it should keep the appliance working steadily to produce hot water at the required rate, or in response to demand.

Solid-fuel appliances may have separate controls to increase the air going up the chimney during initial lighting-up. In this period a rather wasteful amount of heat will go up the chimney, but it is necessary to warm up the chimney so that it will 'draw'. Once the chimney is warm it will stay warm, and thenceforth only the minimum flow of air up it is necessary; the air inlet control will do all that is required.

The above paragraphs explain in very simple terms a most important distinction between gas or oil boilers and solid-fuel fired ones. The former, on/off, type responds to demand much of the time by 'idling' between a 'cut-out' temperature and a rather lower 'cut-in' temperature. Solid-fuel appliances, on the other hand, can be finely tuned to a required output, although they cannot be cut out completely, and relighting is laborious. Appliances with on/off operation, where the controlling thermostat is in the boiler as in the simplest installations, will cut out when the selected water temperature at the boiler is reached. Then the whole system will cool off until water in the boiler reaches the lower 'cut-in' temperature, whereupon the flame will start burning again. Each time the boiler cuts out, heat stored in it and in its casing and chimney is 'wasted', in that it does not get into the water where you want it. The more often this happens, the less efficiently the system will be working, in terms of useful heat supplied for fuel burned.

So such systems are at their most efficient when most is being demanded of them. If you grasp this principle you will understand the basic secret of using them efficiently, because there is a lot you can do to create the right conditions.

The 'cut-in' and 'cut-out' temperatures can usually be adjusted (a professional job), and the further apart they are the less frequently the system will 'cycle'. The effect of this wide temperature gap on the heating system will be that the radiator temperatures will fluctuate to a greater extent. A greater fluctuation will be tolerable in a house with a high thermal capacity, as the heat stored in the structure will tend to even out fluctuations in the supply of heat from the radiators.

Now, the colder the weather is, the more heat you will want from your system. In relatively mild winter weather you have a choice of heating strategies, between heating continuously or near-continuously at relatively low radiator temperatures, and heating in bursts at higher radiator temperatures. The latter strategy will mean the boiler is working harder for less time overall, and ticking over in its least efficient mode of operation for relatively little time. Again, your scope for this technique will be greater in a house that itself stores a lot of heat. You may well find that in some conditions you can get all the comfort you need by putting heat into the house during only two periods during the 24 hours, so the whole process can be governed by an ordinary programmer that gives you two heating periods per day. In colder weather it may suffice to override the programmer at any time during the programmed 'non-heating' period when the house feels chilly.

The effect of frequent 'cycling' on the boiler's efficiency is thought to be considerably less for a modern boiler with a 'low thermal capacity' — i.e. one in which less heat is stored in the heat exchanger and the water, and thus less is wasted as the system cools off each time the boiler cuts out. A lot of factors are involved, however, and boiler performance is very difficult to test in ways that give a conclusive indication of performance in real conditions. The larger the boiler is in relation to the loads on it, the less efficiently it will burn over the course of a season. Many boilers are in fact rather too big, having been installed under a 'guaranteed warmth' agreement. Such boilers will offer more scope for fuel savings if you use control techniques like the one described.

'Whole-house' thermostats

Many central-heating systems, either whole-house or part-house, include a wall-mounted thermostat away from the boiler. This responds to the temperature achieved at that place and switches the system off or on accordingly. Since it reads temperature achieved in the house it should not need to be adjusted in relation to outside temperature, if the distribution system is properly balanced. With such a control the boiler thermostat, if there is one, can be set to any temperature (provided it is high enough to allow the radiators to heat the house up to the temperature called for by the wall thermostat). Warm-air systems have such controls also.

This control works well enough, and makes sure you don't overheat the house as a whole. The balance of temperatures between rooms is determined by the design of the system, or by manual alterations such as the adjustment of radiator valves or warm-air registers (the flaps on warm-air outlets).

The main snag with a wall thermostat is that it reads temperature at one point in the house, while individual room temperatures vary as their gains and losses vary. Normally the temperature you most want to control accurately is that of the living room, but that is not necessarily the best location for the 'stat.

The best place is probably the hall if you have one, and provided it is not draughty. If the 'stat is in the living room you are likely to want to turn it up for 'sitting' comfort at times, in which case the whole house will get warmer to the same extent, which is rather wasteful. 'Supplementary' heating in the living room can provide a higher temperature there at such times, but not elsewhere. A 'stat in the living room would then respond and switch off the heating to the rest of the house, which may not be what you want.

Supplementary heating in the living room is desirable unless you want the same balance of temperatures throughout the house all day, which most people probably do not. If your supplementary heating is an electric fire it may in fact cost less to provide this boost in the one room for a spell than to 'overheat' the whole house by the cheaper fuel. You can of course adjust temperatures by radiator valves or warm-air registers, but such controls are very crude, and don't make for economical comfort. In fact, any sort of 'whole-house' control over heating is bound to be limited in this way, and it is easy to understand how much more desirable room-by-room controls are.

Room controls on appliances and emitters

Room controls have three very great advantages over whole-system controls. First, they allow you to keep each room at the required temperature and no higher (the 'required temperature' for an unused room might be very low), and give you a saving in heat that is likely to be realised in fuel saving. Secondly, they respond to incidental heat gains in that room, so that, for example, sunshine can partially or sometimes entirely heat the room and thus save you heat. Thirdly, they make sure insulation improvements are fully realised, whereas otherwise they are less likely to be.

There is an attendant disadvantage, in that if you leave a window or door open in such a room, and forget about it, the heating in that room will turn itself up to compensate, and you will not notice the difference until the bill comes in. So with room-controlled heating you need more awareness by all members of the household, or some more foolproof method of ventilation control than that of opening windows.

Nevertheless there are strong arguments for room control, particularly of independent appliances. Without it, room heaters frequently have no control at all. Even where they do, there is a human tendency to let the temperature rise as the walls, furniture etc. warm up, when in fact a lower air temperature would be acceptable.

Thermostatic radiator valves (TRVs) are gaining in popularity, and are extremely worthwhile if they are working properly. There are reports, however, that some makes are particularly unreliable in that the chance of finding yourself with a faulty one is unreasonably high. At present you cannot test them before installation, and changing any one after they are installed involves draining the system. You also have to persuade the installer or supplier of the fault. Indeed you may not notice a fault without a concerted effort to test the valves' performance in use. Proceed as follows: when the system is first turned on, with all rooms cold and all valves set at any medium setting, all the radiators should heat up straight away. As soon as that room has reached the appropriate temperature, the valve should shut that radiator down or off so that it cools perceptibly. If you turn it down one number the radiator should certainly shut off for a spell. Thereafter the radiator should either stay at a temperature below its hottest, or alternate between cold and hot, to keep the room at or near a steady temperature. Unless a valve is marked in degrees (which is rare), it does not matter whether the same setting on each valve represents the same room temperature (as long as they all give a reasonable range). In fact this is unlikely, because characteristics of the room will affect its performance in practice. TRVs are vulnerable to silt in the system, so it should be drained and flushed periodically.

Time controls

Time controls are worthwhile for most households. They involve an electric 24-hour clock, which needs to be set and regularly checked for accuracy against any other clock. The clock carries a series of markers — usually four, two marked 'on' and two 'off'. You simply distribute these where you want them around the face of the clock; the whole system will then switch on and off at those times during the 24-hour cycle. Usually with a four-point clock it is also possible to have only one 'on' and one 'off' switching per 24 hours.

Figure 4.3 shows a typical 'programmer', which is a time-clock together with attendant switchgear to connect it to heating and/or hot water according to the needs of summer and winter.

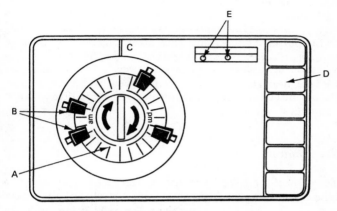

A 24-hour clock (read against mark C).

B Movable 'on' and 'off' switches act as they pass C.

D Push-button switches set the full range of combinations of 'constant', 'timed' and 'off' for space heating and water heating.

E Indicator lights show when space heating and water heating are operating.

Fig. 4.3 Typical 'programmer'

The secret of getting the best from a time switch for heating is to judge the heating-up and cooling-off times appropriate to when you want heat, and how long it takes the house to warm up and cool down. This will vary with outside temperature, for any particular house. Heating up a house from cold is like filling up a bucket with a hole in it. The thermal capacity of the house (see Chapter 3) is represented by the size of the bucket, the coldness of the weather and the house's total rate of heat loss is represented by the size of the hole, and the power of the heating system is represented by the flow from the tap.

Hence in really cold weather it may not pay to have the system cut out at all, if it will not bring the house back to an acceptable temperature by the time you next need it. In milder weather a couple of separate bursts of heat during the day may suffice, even if you are using the house all day.

'Optimum start' controls are made for non-domestic buildings, to time automatically the switching-on of heat according to the outside temperature. They are at present too expensive to be worthwhile for ordinary-sized houses, but cheaper ones will probably appear in due course. However, they do not appear to take into account outside windspeed, which has a more significant effect on the performance of a house than of larger buildings.

Zone controls

Zone controls allow you to supply heat to all or just part of a house at a time, but automatic ones are relatively expensive and usually involve 'motorised' valves, which are prone to failure. Manual zone controls may have a future, however. In mild weather sufficient heat percolates upstairs in a well insulated house to warm the bedrooms, so at such times downstairs radiators alone should suffice. The effect of zone control can of course be achieved by turning off upstairs radiators at their own individual valves in mild weather, and TRVs ought to do a better job than either.

Domestic hot water

The systems in common use for domestic hot water supply are:
• domestic hot water heated by a central-heating boiler (or room heater back-boiler) and stored in a tank sufficient for at least one good bath;
• a domestic hot water storage tank heated by other means — most commonly electricity, but possibly a separate gas heater or solid-fuel stove;

● instantaneous domestic hot water supply, gas or electric, of single or multi-point type.

Solar hot water systems are not yet very common, but are discussed in Chapter 10.

Domestic hot water from central-heating boiler

This is probably the commonest pattern for recent installations. It involves compromising between certain conflicting requirements of such systems, such as:
● the best temperature setting for water heating is unlikely to be best for space heating;
● the boiler must be of a size adequate for both, which is unlikely to produce the most efficient performance throughout the year;
● the circulation of hot water needs to be arranged so as to satisfy the demands of both purposes.

In most cases systems can be improved by more sophisticated controls.

The boiler temperature setting needed for the maximum space-heating requirement is likely to be too hot for domestic hot water: at 180°F there is a danger of scalding at the tap and possibly formation of scale in the hot water cylinder, and at 140°F (central heating in mild weather) the hot water would not be hot enough for some clothes washing requirements. Separate thermostats at the hot water cylinder and at each radiator solve these problems, but at considerable expense.

A large boiler, such as would be chosen for a large or poorly insulated house, will provide adequate domestic hot water with little impact on space-heating output, but will be grossly over-sized and hence rather inefficient during the summer, with domestic hot water only to provide for. The smaller boiler appropriate to a smaller or well-insulated house will be more efficient during the summer, but its radiators will cool perceptibly when all the contents of the domestic hot water cylinder are suddenly drawn off. In those circumstances the domestic hot water cylinder will take some time to recover its temperature, depending upon which of the two demands is given preference within the system.

If the domestic hot water cylinder is included in the pumped circuit (the expression 'pumped primary' describes this arrangement), the order of preference is predetermined. The determining factor is whether the cylinder is heated by the hottest water straight from the boiler, or the coolest – i.e. after its journey round the radiators. In the latter case the cylinder gets the last priority, in the former it gets the first, and the system will behave accordingly.

Perhaps more commonly the supply of heat to the cylinder is provided by gravity (convection currents in large pipes), and is unaffected by the pump. With this arrangement the cylinder gets a modest share of the heat that is available, all the time, except when the pump is switched off, in which case it gets it all, within the limits of how fast the gravity system can feed heat to the cylinder.

Some recent systems with small boilers incorporate a 'diverter valve', which allows you to choose whether the boiler favours the hot water cylinder or the central heating circuit. If you want to provide for a quick succession of baths you can thus do so, at the expense of the house getting cooler for a while. The other alternative is to use an electric immersion heater to 'top up' the temperature of the cylinder – which it will do very rapidly. The same facility will serve if, having no other house temperature control than a boiler thermostat, you want really hot water in the cylinder but not very hot radiators.

Another variation to consider is that of a larger domestic hot water cylinder than the 150 litre (30 gallon) one usually installed in houses. This allows you more baths in quick succession without waiting a long while for the cylinder to recover. But it is then essential to have a diverter valve, and to consider the provision of domestic hot water and space heating as two separate tasks to be done by the boiler, at appropriate times of the day but never simultaneously.

Such a cylinder would need to be extremely well insulated: the conventional '3 inch jacket' would not do. If you left the task of heating up the cylinder to the night-time, you would have to try to ensure that the boiler did not cycle wastefully for a long period after completing this task. The appropriate control no doubt exists but may take some trouble to obtain. With such an arrangement summer hot water production ought to be both more convenient (more hot water available at once) and more economical in fuel, since the boiler would burn continuously until the cylinder were full of hot water, and then shut down until the process were re-activated by some positive decision of yours.

Storage tank heated by separate means

The commonest form of this arrangement is the immersion heater. At its best it provides hot water by a very efficient means but with a very expensive fuel. It is crucial for the storage tank to be very well insulated, because the heat loss from it will be directly related to the temperature of the water inside.

It follows that if your needs for hot water can be concentrated in one part of the day you can allow the cylinder to go cool for the rest of the time. An improvement is to have an immersion heater with two elements, a short one to heat the top of the cylinder for kitchen and washbasin requirements, and a long one to heat the whole cylinder only when you want a bath or to wash clothes. The effectiveness of the arrangement depends on the hottest water staying at the top and not being too much stirred up when water is drawn off.

If you install an extra-large cylinder you may be able to make use of off-peak electricity. It is advisable to negotiate with your Electricity Board before changing to such a system, and decide whether it is worthwhile. You must decide whether you want the option of using some on-peak power occasionally to top-up the heat in the cylinder during the day. This will affect what meter you use.

The traditional 'Ideal' solid-fuel boiler is still probably quite common, although many have been taken out in recent years, in favour of more modern arrangements. Those who still have one will know how its running cost compares today with electricity or a gas 'circulator' boiler. Such boilers give out rather a lot of heat into the room where they stand, which is uncomfortable in the summer. On the other hand, fuel can be eked out by burning some household refuse, or even logs, given the skill and experience to avoid putting the boiler out or being prosecuted under the Clean Air Act.

Gas circulators are simply small boilers, designed for the relatively intermittent use of keeping a cylinder of water hot. Some modern lightweight boilers used for central heating are indistinguishable from circulators. If you have one and are considering getting rid of it as part of the installation of gas central heating, make sure you are making the right decision. For example, could the plumbing be arranged so that you use the circulator for domestic hot water in the summer and the central heating boiler in the winter? Or would you in fact use fuel more economically if you kept the two systems entirely separate? If you heat water at present with an electric immersion heater, might it easily be replaced by a gas circulator, for less running cost?

Instantaneous domestic hot water supply

The gas 'geyser' was one of the first common forms of domestic hot water supply, and plenty are still in use. Their advantage was always that only one (cold) water supply pipe was needed, from the main to the point of use, and when they were introduced gas pipes already existed throughout most houses, for lighting, so gas supply was no problem. Small geysers did not need a flue. The very much larger ones used in 'multi-point' installations did need one (or do now). The latter, while saving on gas pipework and on the cost of individual gas heat exchangers for each point of use, are nevertheless a rather inefficient way of providing hot water. An enormous heat exchanger is needed to provide an instantaneously heated flow of water large enough to fill a bath reasonably quickly, and that whole heat exchanger will light up each time a washbasin tap is used, with a consequently large loss of energy to the flue.

Electric instantaneous water heaters are relatively new. They are a very cheap source of hot water, in capital cost (about £20 to £30 to equip a shower), and very suitable for remote installations, where the heat losses from long runs of pipework would be a major disadvantage of a central supply heated by a cheaper fuel. They are, however, expensive to run, and if they are to be used a lot it is worth the trouble of trying to make a fair estimate of the comparative capital and running costs of the alternatives. Bear in mind that gas-fired hot water is usually produced fairly inefficiently (40—50 per cent) in the summer, depending on how effectively you can contrive to use the boiler (or circulator) in its most efficient way.

General tips on hot water economy

Domestic hot water is inevitably one of the home's most inefficient uses of energy. The main reason is that most of it goes down the drain. But a great deal of the heat is lost before it ever reaches the tap, either through inefficiency at the boiler or from the cylinder or lengths of pipework. A modern installation should be designed to be as compact as possible; older ones, fitted in a house not designed with this factor in mind, are more likely to have significant pipework losses.

It is worth insulating every hot water pipe you can get at (as well as the cylinder, which you should always do first), but the crucial pipes are those that are used often. Every time you run a hot tap, waiting for the hot water to arrive, you are running off water which you have already paid to heat but which has cooled down since the last time the tap was used. Pipe insulation will not guarantee that you never waste heat from that pipe again, but it will mean that on many occasions you avoid it. Filling a kettle from a hot tap, especially if you have waited for the water to run hot, is particularly wasteful, though it does of course save time. While the kettle is boiling, another pipeful of hot water is left behind to cool off. A 6 metre length of 15 mm pipe holds about a litre of water.

Note that with a solid-fuel boiler it may be necessary to have some cylinder insulation removable in summer, to prevent boiling (consult a local expert).

If you are filling a bowl, basin, sink or bath, start with the 'cold' water from the hot tap. This will still be warmer than that from the cold tap, as it will not be below room temperature, and the chances are you will need some cold water in the bowl in any case, to produce a bearable temperature.

A shower is inherently more economical than a bath, as it offers you the chance to get thoroughly clean with about a third or half the quantity of hot water. However, you should realise that if you install a hand-shower in the bath, some people will still wallow in the bath until it fills up, and thus use no less hot water. A stand-up shower with a curtain, even in the bath, probably has a greater chance of success, although of course it costs more.

Studies are being made of the possibilities of reclaiming some of the heat from waste hot water. It will be several years before they lead to a practical and effective technique. The result when they do, however, is likely to be a major step in household energy saving.

5 Routine Savings

This chapter explores possibilities for making fuel savings at no cost. Some of these were established habits a few decades ago, but have been largely forgotten in the age of push-button central heating. Others arise from modern theoretical knowledge, or involve the better use of modern appliances. Such techniques include:

● making the maximum use of sunshine throughout the heating season;

● studiously avoiding over-ventilation;

● choosing your use of rooms to make the greatest possible use of 'incidental heat';

● applying an understanding of comfort so as to be comfortable at lower energy cost;

● getting to know the outside temperature above which no space heating is needed;

● exploiting the 'thermal mass' of the house;

● exploiting the buoyancy of warm air;

● using appliances in a more efficient way;

● making full use of curtains to reduce heat loss from windows;

● making the most use of heat from cooking;

● pre-heating the cold water supply from ambient air to reduce hot water costs.

Some of this advice repeats in condensed form advice given in theoretical terms in other chapters. However, general advice about lighting and electrical appliances is restricted to Chapter 6.

Using sunshine

You will be aware that even in mid-winter the sun can supply some useful heat to a house, but most modern houses are not designed or oriented to exploit it. Heating systems, too, are commonly designed to ignore solar gains, with the result that rooms into which the sun shines can overheat on sunny days unless you juggle with manual controls, or open a window to spill some heat (which often seems easier), or even draw the curtains. The former may be unnecessary or it may not. Both of the latter actions are needlessly wasteful.

The following paragraphs suggest how with a little thought and the minimum of effort you can make full use of such sunshine as you are lucky enough to get. They may make little impact on your fuel costs, but they could make an appreciable difference.

Open the curtains of sunny rooms as early as possible in the day, and open them as widely as possible. Do not open the windows — that will let the heat escape. Even if you are not using such a room, open its door to let the heat percolate into

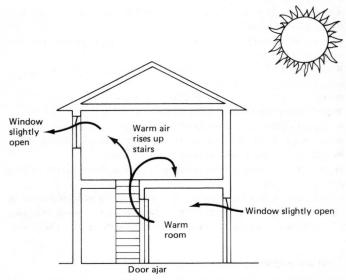

Window slightly open

Warm air rises up stairs

Window slightly open

Warm room

Door ajar

Fig. 5.1 Use of buoyant air movement to draw solar warmth through the house

the rest of the house. If there is a room heater or heat emitter (radiator or warm-air grille), shut it off while the sunshine lasts. Even if the room overheats a little, don't open windows or close curtains if you can avoid it. That way more heat will be stored up for later in the day when the sun has moved round or gone down.

Use only a minimum of ventilation throughout the house. The exception might be if, by ventilation, you can create a slight through draught to carry sun-warmed air through the house and distribute it (*Figure 5.1*). A first-floor window opened slightly on the shady side of the house may help, depending on wind conditions that day. If possible, transfer activities to sunny rooms, and if possible do without central heating. But remember that if you let the sunless parts of the house go cold, it will take time to warm them up again. Get to know what cunning you can employ with your own house to be comfortable without heating as often as possible. In the author's own house, sunshine provides all the warmth needed in early March, with an outside temperature of 7°C (45°F), until about 4.0 p.m.

Avoiding over-ventilation

With ventilation you have to strike a balance between insufficient air, leading to stuffiness and possibly condensation on walls, floors and ceilings, and over-ventilation, which is very wasteful of heat. Research has shown that British people tend to over-ventilate their houses in mild winter weather, when draughts are not so noticeable. If you habitually do this you will use a lot more heating fuel than necessary. One reason is that most windows when they are open at all let through far more air than is needed. If the outside temperature is as high as 13°C (55°F) you should not need heating at all, especially if there is any sunshine. Apart from sunshine, other incidental gains may well provide all the warmth you need to achieve comfort temperatures inside, if you do not waste it through ventilation.

If it is windy outside you probably need no windows open at all. Most houses have sufficient leakage, through other routes than the cracks you may be able to block up, to ventilate them adequately in windy weather.

If you study the passages in Chapter 3 on condensation, you should understand the matter well enough to be able safely to reduce ventilation rates to those you need and no more. You must, of course, provide enough air to feed any fuel-burning appliance and remove pollution: failure to do so can be dangerous. Study also the passages in Chapter 7 on ventilation control, and understand the principle of controllable 'trickle' ventilation.

Remember that warm air rises, and windows left open upstairs will spill out vast quantities of valuable warm air — often causing cold draughts downstairs as well (see *Figure 3.7*).

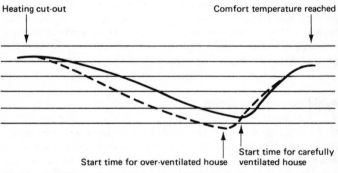

Fig. 5.2 Avoiding over-ventilation permits delaying start time for heating

Unless you have achieved a means of 'trickle ventilation' (Chapter 7), give thought to how you ventilate the house at night and when you go out, in order to achieve the same effect. There should be different habits according to how windy it is; in all cases provide just enough ventilation to freshen the air over the whole period, but if possible avoid scouring all the heat out. By this means you should be able to delay the starting-up time for central heating the next time you need it (*Figure 5.2*).

Never stand talking to someone in an open doorway, either yours or theirs, if you can avoid it. Unless you are afraid of being attacked, bring people into the hall to talk.

Using naturally warm rooms

Many houses have some rooms that receive more than their fair share of incidental heat gains. The kitchen is usually one of these, especially if it has a fridge and a central heating boiler in it. Consider whether it is possible to make more use of this heat by using such rooms more and not heating the rest of the house so much. A corner of a kitchen that is clean, well-decorated and attractively lit might easily double as a living room or homework room at such times. Many people have always arranged their lives around the warm places, so such advice may seem fatuous to them. But others, with full central heating, have come to see rooms as having distinct functions, and so use each one strictly for that function and add heat and light as necessary, because it is so easy to do so. There may be economies to be made from re-thinking such habits.

Understanding the nature of comfort

Refer to Chapter 3 on the nature of comfort, and bear in mind how to apply it. Most of us have become so used to simply turning up the heat in a room if we feel chilly in it that we do not consider whether we could be equally comfortable by another means without spending more on heating. Ask yourself whether you could improve the economy of using that room by:
● eliminating or avoiding a draught;
● improving where you sit in relation to radiant warmth, or keeping away from the coldest surfaces;
● reducing the tendency for warm air to rise to the ceiling and stay there — encourage it to circulate and mix by using a

radiator shelf (*Figure 5.3*) or placing a small fan to push or pull warm air down from high level;
● preventing cold air pooling on the floor, avoiding sitting with your feet in it, or taking special measures to keep your feet warm;

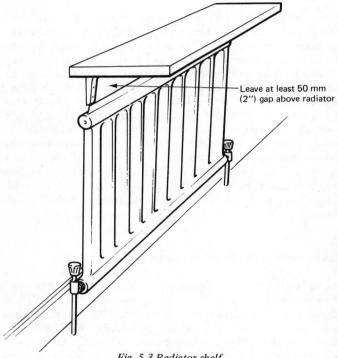

Leave at least 50 mm (2″) gap above radiator

Fig. 5.3 Radiator shelf

● using items of furniture that maximise thermal comfort — for example, a dining chair with a blanket or jacket draped over the back for long hours of sedentary work, as protection against draughts or cold radiation; arranging furniture that people can snuggle into like cats do (who know all about comfort);

- shutting doors to contain the incidental heat gains from lights and/or a television set;
- making sure everyone in the room is equally well placed, or those most sensitive to cold sit in the warmest places.

Getting to know the 'no-heat' temperature

Keep an eye on the outside temperatures in mild winter weather, and get to know at what outside temperatures, in various conditions (of wind, sunshine and internal heat gains) you can be comfortable inside without adding extra heat. In some circumstances it may be a lot more economical to provide just a little heat in one room than to heat the whole house, even if you heat that one room with a relatively expensive fuel.

Exploiting the house's 'thermal mass'

Study Chapter 3 on what this means, and Chapter 4 on the uses of heating systems and controls in relation to the house's thermal mass. If you only want a bit of warmth for a short period, it may be sensible in a 'massive' house to use a 'local' heat source rather than central heating if you have it, and to avoid having to heat up the whole structure.

If you have free heat from sunshine, realise that heat will stay in the room for a long while after the rays of the sun have gone, in a heavy structure. In such a room, even when the sun's heat is no longer enough for comfort, less extra heat will be needed in the evening than in rooms that have been cold all day.

At the end of the evening, switch off heating as early as possible before retiring — get to know how long it takes before the room becomes perceptibly chilly, at various outside temperatures and windspeeds. Make manual adjustments to heating at or before bedtime each night, to avoid leaving an extravagantly heated living room behind when you go to bed.

Exploiting the buoyancy of warm air

Use the fact that warm air rises to help distribute solar heat around the house, if you can usefully do so. In the evening, use the buoyancy of the warm air of living rooms to boost the temperature of bedrooms at bedtime: open the living-room doors at bedtime for this purpose. Conversely, remember that you can heat downstairs rooms more economically if you shut their doors: there is no point in heating unoccupied bedrooms, except of course to warm them up before using them, or when they are in use during waking hours.

Using appliances in the most efficient way

Thoroughly get to know your own heating system: Chapter 4 should help. There is usually a great deal to be learned about any system in order to get the best efficiency from it. This means more comfort per pound spent on fuel, and hence (usually) less money spent over the course of the year.

Making full use of curtains

Curtained windows, particularly if the curtains are close fitting and well lined, can be as good as or better than double-glazing at retaining heat. See Chapter 8 on what makes for efficient curtaining. During the heating season, close all curtains as soon as it gets dark − or even sooner if you can bear it and there is no more sunshine around.

Some curtain tracks stand clear of the wall to which they are fixed. If possible fix hardboard or stiff card neatly across the top to reduce the flow of warm air down the face of the window (*Figure 5.4*). The traditional 'box pelmet' does this job admirably, and prevents air currents from causing warmth to be 'pumped' out of the room in this way.

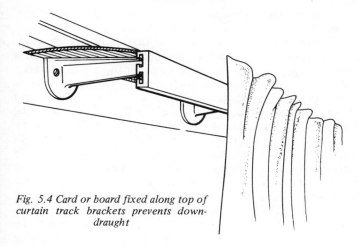

Fig. 5.4 Card or board fixed along top of curtain track brackets prevents down-draught

Heat used in cooking

Some understanding of the heat output of the cooker is advisable, for three reasons:
● safety;
● having understood the safety aspect, you may be able to use heat from the cooker as a source of space heating, either regularly or in an emergency;
● you may be motivated to use it more economically.

It is necessary to consider gas and electric cookers separately, because they have important differences. A gas cooker has the same order of heat output as a central-heating system. For example, the author's own central-heating/hot-water boiler burns gas at about 50 500 BTU/hour and the gas cooker (with the oven, four rings and a grill all on) burns about 53 700 BTU/hour (0.51 and 0.54 therms/hour or 14.8 and 15.7 kW, respectively). However, the central-heating boiler has a 'rated output' of 38 000 BTU/hour (i.e. it delivers this much useful heat), which is about 75 per cent of its rate of burning gas. Against this, the cooker delivers *all* the heat value of the gas it burns into the house, because it has no flue. So the cooker is a source of heat one-third more powerful, through being about 100 per cent efficient (at least as a producer of heat in the house).

It does seem rather illogical that the central-heating boiler has to have a flue while the cooker does not, especially as this flue disposes of a quarter of the power of the boiler. However, the boiler may be running continuously, while the cooker is only expected to be used for an hour or two at a time, and during waking hours. The room where the cooker is must be adequately ventilated, of course (as is the case for a central-heating boiler, except one with a balanced flue). But such ventilation does not prevent all the waste gases staying in the house. In fact it may well assist, because a strong current of hot air from the cooker, backed by a source of cool air from outside, is quite likely to pour into the rest of the house by 'stack effect', unless barred by a closed door.

Certain observations follow. Firstly, the cooker provides by far the biggest incidental heat gain in the house: even one large burner represents some 17 per cent of the cooker's output, equivalent to 9000 BTU/hour or about 2½ kW. Up to a point you probably have the choice as to whether to use this heat or forcefully to reject it, e.g. by a cooker hood or extract fan. To use it all would probably be unwise, as there may be potential lung irritants in the burned gases; in any case these gases usually carry dirt, grease and smell, and almost certainly water vapour. Such health risks as there are, are small: burned natural gas consists largely of carbon dioxide and water, but also traces of oxides of nitrogen, which, in quantity, can be harmful to lung sufferers. Some people of limited means consciously use gas cookers for space heating, perhaps without realising that, leaving all other considerations aside, they are using their gas in the most efficient way possible.

Given that the cooker provides so much heat, it also follows that it will pay to use it as efficiently as possible for cooking. This matters most outside the heating season, when the waste heat is of no other benefit. If, however, your heating system and the layout of the house do not make it possible to exploit the waste heat to help warm the house, then efficient use of the cooker is doubly important.

Gas cookers are limited in their scope for more efficient cooking. The use of broad-bottomed pans with lids, and especially of pressure cookers, no doubt helps. A gas overhead grill at about 3¼ kW is a very inefficient way of cooking — especially while it is warming up (an electric toaster of 1¾ kW will make toast a lot faster, but the fuel costs about five times as much as the lower gas tariff). It is certainly worthwhile, too, to cook as many dishes as possible simultaneously in the oven, which is in itself the most efficient part of the cooker (gas or electric) because of its retained heat and thermostatic control.

Electric cooking is inherently less wasteful of heat, partly because there are no burned gases to vent off, and partly because, where boiling rings are concerned, flat solid plates in contact with the bottoms of pans and kettles inevitably put more of the heat where it is required. The maximum heat output from electric cookers is typically up to 7 kW.

There is probably scope for better insulation of ovens of all types; present insulation standards are designed to keep outside surface temperatures within safe limits rather than to conserve heat for its own sake. The gas-cooker owner can experiment to find out which common processes normally done over gas might be done more economically by purpose-built electrical appliances. The improved cooking efficiency does not, however, outweigh the present differences in tariff. Typical annual consumptions for families are 60–80 therms of gas (£14–£18 at the *higher* rate of 22.8p/therm), as against 1000–1500 kW h of electricity (£29–£43 at 2.9p/ kW h).

The electric kettle is an exceedingly efficient way of boiling water, being both direct and very fast. The electric slow cooker, however, loses a great deal of heat simply because it takes so long.

Helping water heating with ambient energy in summer

The temperature of cold water entering the house through the main (normally well below outside air temperature) will usually vary through the seasons, but also depends on the source from which the water is drawn. The warmer the supply to the hot water system, the less your hot water will cost. If, as is usual, the water stands in a large tank in the loft before it is used, it will warm up to something approaching air temperature, especially if the loftspace is heated by the sun. Hence, for example, if by this means you can warm the cold water supply from $10°C$ to $15°C$, it will take 12½ per cent less heat to raise its temperature to $50°C$ in the hot water system. Such a saving is unlikely to be dramatic, but would nevertheless justify removing the insulation from the cold-water tank at the start of the summer. Don't forget to put it back when the heating season starts again.

None of the suggestions listed in this chapter is likely to make on its own a large difference to your consumption, but the sum of all of them might easily equal the yield from a fairly expensive modification to the house, whereas most of them will have cost nothing at all.

6 Electric Lighting and Appliances

Chapters 1 and 2 stressed the cost of electric consumption in comparison with other annual fuel costs, and urged you to try to establish how your annual consumption is distributed between uses. Lifestyles differ a great deal, and what is true for one family may be quite false for another. But the fact remains that every unit of electricity costs little under 3p, at the time of writing, and much of the cost is continous round the year, so there is real money to be saved by making one's uses of it more efficient.

The immediate reaction to this sort of statement may be to say 'but I don't use more lighting than I need, and I can't insulate a washing machine'. However, there are ways of using lighting that can provide as much or even more useful illumination from less electricity, and when the time comes to replace a household appliance it may well be possible to buy one that does the same job more efficiently.

Lighting can account for a surprisingly large part of your annual electricity consumption. (For example, in the author's own house the proportion is about two-thirds: this sounds profligate, but it arises partly because the family disperses to different rooms most evenings, and partly because of the nature of the house, which has a dark internal staircase that has to be well lit all day except in very bright weather.) There is plenty of scope for saving, simply because much of the lighting uses far more power than is needed to achieve the desired effect.

Efficient use of lighting

Ordinary household lighting is inherently inefficient at present, largely because the common light-bulb (tungsten filament) burns one watt of electricity for a light output of 12 'lumens' (on average). The lumen is the unit of total light available from a light source. For comparison, some light sources available for industrial purposes give as much as 135 lumens per watt, and so are more than ten times as efficient. The arguments in favour of tungsten lamps are their cheapness and adaptability, and their acceptable colour for domestic purposes. It can only be a matter of time before a more efficient but equally acceptable bulb is available for domestic use, whose extra cost will be justified by the electricity it saves. Basically, the less efficient lamps turn more of the power they consume into heat — useful in winter, but there are cheaper ways of heating a house.

Fluorescent lights are already familiar in homes, and these produce up to 125 lumens per watt, depending on colour. Unfortunately the few colours thought to be acceptable for domestic use are some of the least efficient, but 40 lumens per watt is a useful working figure. Even this is 3½ times the efficiency of tungsten lighting, so there is a very good argument for using fluorescent tubes wherever possible. They usually present no problem in kitchens and bathrooms, but can also be used elsewhere, if incorporated into a fitting that makes their kind of light compatible with domestic decor.

The longer a light has to stay on in the course of a day, the stronger is the case for trying the fluorescent type. For example, the staircase lights mentioned earlier probably burn for eight hours a day on average round the year, and these need to be two 60-watt bulbs, consuming some 350 kW h annually. The same job done with a 40-watt fluorescent fitting would use 120 kW h. The saving would be worth nearly £7 a year. Many people will save no more than that from roof insulation, and the fluorescent fitting would probably cost less (ignoring the government grant available for roof insulation).

Certain snags about fluorescent lamps must be realised.

● Replacement tubes cost more each, although the replacement costs over a period of time are probably no greater than for tungsten.

● As the tube gets older it becomes less efficient, down to about 75 per cent of its 'new' efficiency after some 9000 hours' use. (Tungsten lamps typically blow after about 1000 hours or less.) Tubes must be kept clean.

● Some fluorescent fittings buzz; this is said to be curable by a simple electrical modification.

● High-frequency flicker is inevitable; most people can't see it but others can't stand it.

Efficiency and positioning of fittings

Tungsten light-bulbs emit light more or less evenly in all directions from the filament. However, the majority of light fittings, for decorative reasons, absorb a large proportion of the light given off, or direct it towards light-absorbing surfaces, where it neither gives pleasure nor serves any useful function. Consider therefore the following 'rules', and see whether as a result you can reduce the total wattage of your lighting.

● Distinguish between general lighting, 'mood' lighting and 'task' lighting, and see that each kind does its job efficiently.

● Try to use fittings that let out as much as possible of their light, either directly or by means of a reflector (shiny or silvered) that throws out a beam suitable for the purpose.

● If the shade is supposed to be translucent, and serves only to reduce the glare of the bulb or its filament, make sure it really does that, and above all that it is kept clean.

● If a large proportion of the light falls on a wall, ceiling, carpet or furniture, realise what proportion is being absorbed (see 'reflectance factors' below).

● Position or direct lights so that they concentrate brightness where it is needed and not elsewhere. Remember that brightness diminishes with the square of distance; i.e. a light two metres from a surface lights it only a quarter as brightly as if it were one metre away, and at three metres is only one-ninth as bright, and so on.

● If your scheme of lighting involves reflecting light off decorated surfaces, make sure they are clean, and preferably brilliant white.

Reflectance factors

Shades of paint are quoted by manufacturers (if you ask) in terms of their 'Munsell value' or their 'reflectance factor'. These indicate what proportion of white light falling on that surface gets reflected back. Munsell values operate thus:

$$\text{Reflectance (per cent)} = \frac{V(V-1)}{100}$$

For example, if a paint has a Munsell value of 7, its reflectance is

$$\frac{7(7-1)}{100} = \frac{7 \times 6}{100} = 42 \text{ per cent}$$

Most colours reflect back only a small proportion of light, and even white emulsion only yields 80 per cent, with brilliant white just a little better. 'Pastel shades' tend to be around 40–50 per cent, and today's popular strong colours are in the range 10 to 30 per cent. These figures are not offered for calculation purposes, but just to indicate the effect of decor on lighting levels, and the case for lighting more carefully if you have strong or dark colours and do not want them to soak up expensive electricity.

Task lighting

In kitchens, bathrooms, workshops and utility rooms, overall strong lighting is usually needed and it is here that fluorescent lamps come into their own. Other rooms can do with much less light, except where there is a job to be done such as reading, sewing etc. In these places it is far more economical to provide small pools of bright light as and when needed than to try to light the whole area well from the ceiling; many people find this rather clinical in any case, and it is uneconomical if done with tungsten lights, which are often the only sort acceptable.

A really good task light (such as a modern 'Anglepoise') gives excellent lighting to work by and lights the rest of the room as well, unless there is additional task lighting needed. But try to arrange that such task lights do not get used in addition to general lighting. To do so is to use them as a form of decor — a function served by 'mood' lighting, which can satisfactorily be of lower power. Mood lighting serves only to make a room interesting and cosy, and is often done with fittings that do in fact absorb a lot of their own light. But don't try and make them double as task lights, or you will tend to want over-powerful bulbs in them — and maybe even then you will get eyestrain trying to read by them.

Finally, where a task light is needed a lot in one place — such as over a child's homework desk — seriously consider a small fluorescent fitting. A shield can be provided to mask its glare and another (also low-powered) bedside light provided as well, giving a very localised light for reading in bed. In this way the majority of the time will be served by the 'working light' at, say, 20 watts (with a fitting this will appear as bright as a 60-watt tungsten bulb); when that is off, a tungsten reading-lamp of, say, 40 watts can be used. For comparison, a tungsten over-head light of at least 100 watts would be needed to serve both purposes, while still being inadequate for either.

Suitably shielded, small fluorescent fittings might serve most circulation areas where the lights tend to get left on for long periods. Work out an alternative lighting scheme, and see whether it might not justify some investment in more efficient fittings.

Finally, avoid using electric lighting where more daylight could be let in instead. For example, could a dark circulation area be lit by natural light for part of the day by glazing the panels of one or two doors? This can produce a very pleasant effect, as well as saving electricity.

Appliances

The big spenders, apart from cookers and immersion heaters, are freezers and fridges, laundry equipment and dishwashers, and colour television. The following notes are intended as helpful general principles, but there is no substitute for a meter-test as described in Chapter 2.

Television sets have become a great deal more economical over recent years. Early colour sets required 350 watts or more, but many modern ones use around 150 watts. If you rent your set, check (from the plate on the back) how much power it consumes. If you use the set a lot, it may pay to change to a more modern one at a rather higher rental, since at five hours viewing a day a change may save £10—£12 a year.

The following notes are based on German experience and study. (The source gave the impression that Germany is particularly conscious about appliance economy. However, in tests of washing machines and TVs available in Britain, the German ones were among the highest consumers.)

Freezers and refrigerators differ considerably in their consumption between makes and models, largely determined by:
● their shape and surface area (and whether upright or chest-type);
● how well insulated they are;
● the performance of the refrigeration mechanism (heat-exchange system) and its mechanical efficiency;
● the design and construction of door/lid seals.

A German standard has been set down whereby, for freezers, consumption in kW h per day by a standard test should not exceed a given factor related to their volume in litres. (Cabinet and chest freezers are judged separately, since chest freezers consume only 85 per cent of the electricity of cabinet freezers of the same size.) It is perhaps reasonable to assume that new designs of German freezers will be made to these standards. Good examples from a list of German refrigerators and freezers have daily consumptions of about the following:

Floor-model refrigerator, 3-star model	140 litres	1.5 kW h
Floor-model freezer	200 litres	2.6 kW h
Chest-type freezer	200 litres	2.0 kW h
Combined refrigerator/freezer	250 litres (gross)	1.3 kW h
Combined refrigerator/freezer	200/150 litres (gross)	1.6 kW h

It is worth studying reports of consumer tests before buying such a significant item. A 1.5 kW h per day appliance will cost you around £16 a year to run, and a poor performer could cost you double that.

There is considerable difference in consumption between different types of washing machine. Automatics use more than twin-tubs if they heat their own water electrically (especially for the 'boiling wash' programme), because a twin-tub can be re-loaded using the same water two or three times whereas an automatic 'dumps' after each wash cycle. But front-loaded machines use *less* water per wash, and hence less power for washing and spinning. These facts seem to indicate that automatics may well be cheaper to run, provided you use hot water from the domestic hot water system rather than heat it electrically in the machine. All machines are more economical if properly filled, and they don't use appreciably less electricity when under-loaded.

With electric dryers, the consumption seems to reflect almost entirely how much water you remove from the clothes with any given drying technique. In general, spinning is more efficient than heat-drying of any kind, and air-exhausting tumble dryers use less power than the condensing type. It seems that the secret of economy is to spin out as much water as possible, and to avoid running a tumble-dryer longer than absolutely necessary. Some of the latter cut out automatically according to dryness rather than by time-switch, and this is probably a very useful device for achieving economy.

Laundering machinery demonstrates that heat costs a lot more than mechanical power to do the same job. But remember that an extravagant washing machine may operate only once a week, whereas an extravagant refrigerator operates 365 days a year.

Electric cookers differ in the amount of their oven insulation. Insulation standards at present are governed by the safety aspect, i.e. maximum outside casing temperature. For this reason, an oven with pyrolytic cleaning, which can heat up to 500°C, will be extra well insulated, and the resulting saving in consumption will more than outweigh the extra power used for the cleaning process. Air-recirculating ovens are *not* normally less well-insulated than conventional ovens, and so, because they get the same baking results at a lower temperature, they use the least power of all.

Remember, however, that the power consumption of the oven may be quite small in relation to that of the whole cooker, depending on how much it is used.

Dishwashers vary (according to design) in their electricity consumption and in the amount of water they need. The hottest washes use the most power, as also do the machines using heat to dry the dishes. So it seems that running costs will be determined by how fast you decide you want the operation done, at the time you choose your machine. Consider also how often you will use it, and do the arithmetic.

Clearly there are not many subleties in the factors that determine appliance consumption. The really big consumers are refrigerators and freezers, partly owing to a glaring lack of logic. They cool the food inside and release its heat from the back of the cabinet into the room where the cabinet stands, thus all the time adding to their own load. What is needed is a refrigerator (or at least a freezer) whose output heat can be utilised indoors in the winter and released outdoors in the summer, or else used to pre-heat hot water supply. Perhaps someday someone will produce refrigerators and freezers whose (hot) condensing coils are remote from the cabinet, and can be moved between seasons − although it might be cheaper simply to increase the insulation of the cabinets. The heat pump (Chapter 11) is little more than an inside-out refrigerator, so maybe if these ever become commonplace they will be linked to refrigerators, with a consequent power saving by both. Meanwhile, the simplest practical advice is to keep a freezer in a cool outbuilding, and a refrigerator with its back to a cool wall if possible.

7 Insulation (1)

This chapter and the next deal exclusively with physical improvements that you can achieve or organise yourself with little difficulty. We begin with the three areas where normally the most significant improvements can be made: roofs, walls, and ventilation control. Chapter 8 then deals with windows, floors and doors, and a number of other minor but useful possibilities.

If you are making alterations yourself, or instructing a builder to make them, please read Chapter 9 first. You and only you are responsible for the condition and safety of your house, and it is most important to avoid storing up trouble for the future by what you do.

Roof insulation

You will have seen from the whole-house examples in Chapter 3 how a proportion of the whole heat loss can be attributed to each element of a particular house, with given assumptions about temperatures. This should lead you to an estimate of the annual value of heat losses from the roof. The degree of improvement you make to your roof insulation should determine how much you reduce those losses, *provided* that the house temperatures stay the same afterwards. But unless you can control them they will not.

In fact in a large proportion of British houses there is no control over bedroom temperatures, and they are in practice heated most of the time by heat 'escaping' from below. In such houses, unless you can reduce this supply of heat (e.g. by insulating bedroom floors or downstairs ceilings or, more important, reducing heat supplied by buoyant air) inevitably you

will get warmer bedrooms. If it is only the roof you are insulating (rather than walls as well), you will then lose more heat
through the walls and windows of the upper floor (or floors),
in proportion to their higher temperatures. Thus the fuel you
save as a result of the roof insulation may be modest compared
with the theoretical reduction in total house heat loss (see
Figure 3.9).

However, roof insulation *is* worthwhile for most houses,
largely because it is so quick, easy and cheap to achieve (and
the result is permanent). This is all the more so if you can get
two-thirds of the cost covered by a Government grant. On the
other hand, the provisos mentioned above illustrate that there
is a lot more to energy conservation than just putting some
insulation in your loft.

Performance of various materials

A completely uninsulated roof can have a U-value of from
around 3.0 down to about 2.0 W/m² °C. The former will apply
to a slate roof with no sarking (felt or boards laid directly under
the top covering), the latter to one with fairly thick tiles and
felt. Others can lie between these extremes. Starting with a U-
value of 2.2 W/m² °C, the effect of the commonest form of
insulation (glass or mineral fibre) will be as indicated in *Figure
7.1*.

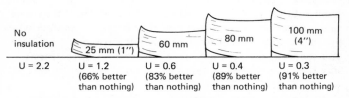

No insulation	25 mm (1″)	60 mm	80 mm	100 mm (4″)
U = 2.2	U = 1.2 (66% better than nothing)	U = 0.6 (83% better than nothing)	U = 0.4 (89% better than nothing)	U = 0.3 (91% better than nothing)

Fig. 7.1 Relative merits of various thicknesses of roof insulation (mineral fibre)

The first thing to notice is how, the more insulation you add,
the less return you get for each extra unit of thickness. Since
at present you usually pay a price directly related to thickness,

the really thick material is seldom worthwhile. It is fairly easy to be misled by the enthusiastic advertising of insulation manufacturers. Here you must make your own decision − although you may be forced into using more than you can justify if the more modest thicknesses are no longer available from retailers. The equivalent insulating values of other materials that might be used for roof insulation are as follows (related to an assumed 'uninsulated' U-value of 1.9 W/m^2 $^{\circ}$C):

80 mm glass-fibre mat	U = 0.4 W/m^2 $^{\circ}$C
80 mm mineral-fibre mat	U = 0.39
80 mm blown mineral fibre	U = 0.37
80 mm cellulose fibre	U = 0.38
130 mm vermiculite	U = 0.39
75 mm expanded polystyrene	U = 0.39

Figures quoted for the U-value of a roof insulated with a particular material are unlikely to be accurate, because the final value will depend on the precise construction beforehand, and upon how it is applied − e.g. over or between the joists. Some materials can be manufactured in differing densities, and their performance will depend on this to some extent, but at present this factor is not normally quoted on the packaging. With mineral-fibre materials, the lower the density, the greater the chance of heat being lost by convection currents acting in the upper layers of a thick quilt. So do not be impressed by quotations of performance that pretend to a high degree of accuracy.

A further possibility is reflecting foil as a backing to tough building paper − see later. This has a much lower effect on U-value than other types of insulation, but is a very useful material for some purposes.

Materials and techniques

The commonest material to date − quilt made of glass or mineral fibre − is simply rolled out either between the joists or over them. It is usually supplied in 400 mm (16 inch) width rolls, which fit nicely between joists spaced at 450 mm (18 inch) intervals. But a lot of older houses have different joist-spacing, and the spaces may be smaller or even irregular. If this is the case you can roll the quilt out *across* the joists, or you can choose a granular or loose fill instead. With the former technique, extra care is needed in walking around in the loft afterwards. Also the insulation will tend in time to sag into the spaces between joists. Laying it across the joists will initially give a slightly better U-value, but of course uses more material.

The term 'granular fill' means polystyrene beads or chips, chips of other foam plastics, and also exfoliated vermiculite ('Micafil'). It should be spread out between joists, preferably up to their depth only. You can lay it deeper, but if you do it is difficult to level off to an even thickness, and also to see the joists so that you can walk safely. 'Loose fill' applies to loose mineral fibre or cellulose fibre, both of which lend themselves to installation by a contractor using a large blower and hosepipe − a very quick and efficient method given certain precautions. The man working the hosepipe must be good at directing the flow evenly and avoiding drifts and 'shadows', and must be careful to avoid blocking future ventilation of the roof space.

Reflective foil is best applied to the underside of rafters, and fixed with a staple gun. Given the material's limited effectiveness, this keeps the whole roof space (or part of it) warm, which is useful if you store things in it. The most practical material is foil-faced building paper, which has a tough waterproof and reinforced paper backing. You can if you wish tape the joints, or decorate it afterwards, for a very cheap 'ceiling'. There is a flame-retardant grade by way of a sensible safety precaution if you are using the space under it. But it costs more, and appears to be available only in 1220 mm (4 ft) × 100 m (333 ft) rolls − an enormous area. (See Appendix 4.)

Precautions

Many materials have certain undesirable properties, which are described in Chapter 9. Also, certain precautions are necessary before you start, as with any material.

● Seal off *all* holes between the house and the roof space, even those where cables pass down to light fittings and switches. Make sure the materials you use will not attack the plastic of the cable sheathing (there is an aerosol foam that seems ideal for this purpose – see Appendix 4).

● Check over your wiring. Old cables may tend to overheat already, but insulation laid over them may cause them to catch fire. Old-fashioned cloth-covered cables may be waxed, and this

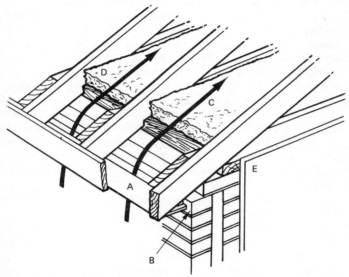

Fig. 7.2 Precautions regarding ventilation of roofspace in 'typical' construction (houses vary considerably). Outside soffit (A) has its inner edge clear of the wall, leaving slot (B) for ventilation. (If it had not, ventilation holes would have been necessary: about two 25 mm (1 in) diameter holes per 450 mm (18 in) joist spacing.) Route must be clear for air to pass up into roofspace(C). Ensure insulation does not block air flow at its outer edge (D), but that it covers the whole width of the ceiling (E); failure in either respect spells trouble, If you have to insulate over joists, cut small squares to put between joists along the outer edges, otherwise you may get condensation and mould along fringes of bedroom ceilings. If it is very difficult to achieve ventilation, special roof-pitch ventilators are available (see Appendix 4)

stuff burns very readily. Make sure junction boxes are intact
and properly closed, that no cables will be stretched by the insulation, and no cables after being covered may later be walked
on.

● Check for roof leaks, and fix them (some can be done from
inside).

● Make sure the whole roofspace is ventilated. This is very important because, after insulation, condensation could occur
there and at worst can cause rot in timbers or the corrosion of
metal fastenings. Roofs with no sarking (where you can see the
underside of tiles or slates) are perfectly safe. Others need some
special ventilation, depending on construction (see *Figure 7.2*).
Don't tuck insulation into the eaves, but do ensure that the
insulation covers all the area of the ceiling below.

● If you are handling glass or mineral fibre, consider protective
clothing. A cheap medical or industrial mouth and nose mask
is the best precaution, as the lungs are most at risk (though the
risk is probably small). It is essential to bath or shower afterwards. Gloves are optional; you may prefer to use bare hands,
and work on a cool day, because the fibres don't seem to bother
cool dry skin, whereas with gloves inevitably your wrist will
sweat. Use a mask against dust with vermiculite.

● Take the trouble to provide powerful, safe lighting for the
job. It saves the time it takes, many times over.

● Take planks to walk over if you feel you need them, but
don't expect to do what some pamphlets suggest and kneel
on a short plank, unrolling the insulation towards you. That way
you will take all day and probably have an accident as well.
If you use a long plank to move quickly from the hatch to
where you are working, avoid stepping on an unsupported end.

Tanks and pipes

After you have insulated the roof at ceiling level, cold water
tanks in the roof will naturally become colder and far more
likely to freeze. Insulate these at the same time — don't put it
off. Make sure insulating materials and their dust are kept out

of the water. Avoid trying to insulate *under* the tanks, even if it looks possible, as the heat leaking up at this point should further help to ensure they don't freeze. It is important to make tank insulation easily removable: you will need to change washers on ball valves from time to time, and if the cold water storage tank can be allowed to warm up in the summer it will make you a useful saving in hot water costs (see Chapter 5).

Pipes also must be lagged. Include any expansion pipes (coming up from a hot water cylinder and possibly from a central-heating radiator system; both are important — see Chapter 9), up to about a foot above water level in the tank. These are the ones that rise next to the tank and turn over to pour into it. Also lag the overflow pipes from the tanks; otherwise, in the rare event of a ball valve failing in freezing weather, you may regret heartily not having done so. The quickest form of pipe insulation to fix is preformed polyurethane foam, but this is the most expensive. It will need some tying on to prevent gaping at joints, especially on bends. Note that sunshine destroys polyurethane foam, so cover it up if you use it near a rooflight. Felt is quite nice to work with, but wrapping it round and tying it on with string is laborious (or you could use old socks or blanket — just as good if two or three layers thick). Glass-fibre pipe wrapping is also very good, but rather unpleasant to work with, especially where the pipe is fixed on a wall or joist: feeding the wrapping through the gap may cause it to fray.

Insulating the pitch of the roof

In some circumstances it may be desirable to insulate the roof at rafter level, rather than over or between the ceiling joists. This takes more material usually, but may be the logical answer for certain awkward roofs, or where you want all or part of the loft space kept warm. The technical problems here are:
- to fix the insulation permanently and satisfactorily;
- to avoid fall-out of possibly irritant fibres;
- to guard against condensation within the material;
- possible difficulty in future in locating roof surface leaks.

The first two are not difficult, but it is almost impossible to guarantee the third, which is in turn incompatible with the fourth.

Condensation is probably inevitable, because on cold clear nights the outer surface of the roof will become very cold (several degrees below air temperature) and any moist air coming into contact with it will form condensation, which will tend to soak into your insulation, rafters and tile battens. The insulation will both promote condensation by making the inside roof surface even colder, and also reduce the chance of it drying out the following morning, which would normally happen without insulation. In theory a sheet of polythene or foil stapled on the inside of the insulation should act as a 'vapour barrier', but experience has shown that it is virtually impossible to form an effective barrier to vapour.

If the roof has no sarking you should have no problem, as the gaps between slates or tiles should adequately ventilate the space. You would be sensible to use foam plastic (e.g. expanded polystyrene) board insulation, leaving as large as possible a gap between it and the tiles, and make sure that the gap is well ventilated. Fitting this material accurately is bound to be laborious, unless you nail it below the rafters. To minimise waste, buy the largest boards you can get into the loft (*Figure 7.3*).

Another method if there is no sarking is to staple mineral-fibre insulation between joists, but again make sure to leave a space above it for ventilation. Polythene stapled underneath will prevent fall-out of fibres and also any tendency of the material to sag over time. The depth of your rafters, less *at least* 25 mm (1 in) for air circulation, limits the thickness you can use to a maximum of about 80 mm (3 in).

If the roof does have sarking, you should leave 50 mm (2 in) ventilation space above, so you must use a board material unless you can get mineral-fibre quilt thin enough. Foam plastics have approximately the same insulation value per unit thickness as mineral fibre, so this thinner board will give you less insulation but more effective airflow between it and the roof surface. See *Figure 7.4*. You may be able to simply press slabs

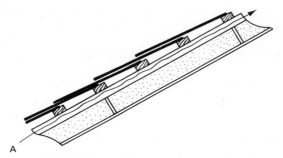

Fig. 7.3 Roof pitch insulation where there is no sarking: 75 mm expanded polystyrene boards jammed between rafters, leaving at least 25 mm gap (A) for ventilation. Remove top and bottom slabs in summer to dry out timber if damp with winter condensation. (Maximum of 80 mm mineral-fibre quilt can be used instead, with sheet polythene fixed on underside to prevent dust fall-out)

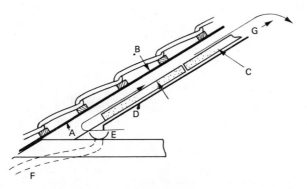

Fig. 7.4 Roof pitch insulation where there is sarking: sarking felt or other sheet (A) prevents ventilation of void over insulation, so gap (B) must be at least 50 mm deep, allowing max. 50 mm insulation (C) if fixed within depth of rafters. Since condensation is bound to occur on underside of slates/ tiles, sheet under with polythene (D), and preferably collect drips in a gutter (E) draining out through eaves (F). Leave gap (G) at top edge of insulation, so air can circulate behind

of it into position between rafters, thus allowing for removal to check on leaks. Also you could then remove a whole (horizontal) row of slabs top and bottom in the summer to make sure timber dries out. If the roof has a fairly steep pitch (30° or steeper) and good overlaps to tiles or slates, it is usually safe to perforate the sarking or open up its joints for ventilation of the space below.

Internal walls on to roofspaces (Figure 7.5)

Vertical walls between rooms and roofspaces can be prime sources of heat loss, especially as they are usually only of minimal thickness. If brick, they are likely to be only 112 mm (4½ in) thick, and if timber and plaster, they may have even

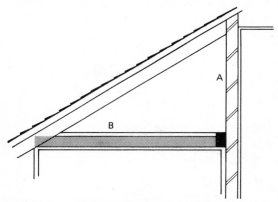

Fig. 7.5 Internal wall adjoining roofspace (typically in split-level houses): both wall (A) and ceiling (B) should be insulated on roofspace side

less resistance to heat loss. It is easiest to insulate these on the cold side. There may be some condensation, but never so much as under roof slopes. Use mineral fibre with timber construction, so that any moisture can easily dry out into the roofspace. Do *not* cover this with an impermeable layer on the roofspace side. If the wall is brick, you could fix mineral fibre between or under battens, but it would be better to use plastic-foam slabs fixed with adhesive.

Flat roofs and 'voidless' roofs (Figure 7.6)

Both these situations are difficult to cope with, and considerably more expensive than ordinary loft insulation, even for a more modest increase in insulation value. The problems to be solved are:

● To maintain the ventilation of the voids between the structural timbers, or better still improve it, because after insulation the timbers may be more subject to condensation. Hence the new material normally has to be fixed up to the existing ceiling.

● To find an insulating material that will give a worthwhile improvement but is not too heavy for safe fixing, and does not constitute a fire risk.

● To finish with an acceptable ceiling surface, hard enough for subsequent redecoration.

In practice the choice at present is between a lightweight material that can be fixed with adhesive and has (or can be given) an adequately robust surface 'skim', and a sandwich of insulation and plasterboard. The latter would provide both surface finish and fire protection, where this is necessary.

There are foam plastic materials for which fire safety can reasonably be claimed, but they are expensive — about £4 per square metre at 25 mm (1 in) thickness. Expanded polystyrene is normally regarded as the cheapest fairly rigid foam insulant, but is frowned on by many fire authorities; 9 mm polystyrene tiles are usually thought acceptable, but their insulation value is quite small (see Appendix 5).

A firm in the North of England recently started to manufacture quite thick (up to 50 mm) polystyrene slabs encased in glass-fibre-reinforced plaster skim (see Appendix 4). This is claimed to have a satisfactory resistance to fire, but fixing is a specialist job and it ends up quite expensive. You could fix up plain polystyrene slabs with adhesive and apply a plaster skim

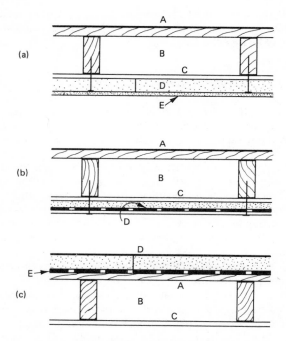

Fig. 7.6 Flat and voidless roofs, with roof deck and weather surface (A) and existing ceiling (C). Three methods of insulation; void (B) must be well ventilated to outside air in methods (a) and (b), but to inside air in method (c)

(a) Expanded polystyrene slabs (D), up to 75 mm thick, with plaster skim coat (E); slabs fixed with adhesive, and preferably nails also. Vapour check not possible

(b) Proprietary 'sandwich' board of insulation and plasterboard, incorporating vapour check (D); plasterboard is screwed to joists above

(c) Where weather surface (D) is being renewed, insulation can be laid over deck, with vapour-check membrane (E) underneath. As air can get under this covering, it should be fixed down against wind (see appropriate trade literature)

coat underneath. Both these methods, however, carry a slight risk of vapour permeating at the joints and causing trouble at the cold surface above, although generous ventilation of the void above the existing plaster should be adequate protection.

Perhaps the method that is safest all round is to fix up a plasterboard/polystyrene sandwich board, using long screws into the joists above. This is hard work because the material is quite heavy, but whole sheets could be cut in half first. Use the sort with an integral polythene 'vapour check', and treat the joints after with scrim and filler before painting.

A final possibility is to wait until the roof — whether pitched or flat — needs major maintenance. The task of stripping it and adding insulation above the rafters (or deck) is then fairly easy and should be only a small proportion of a contractors' price for the job, or of your own labour for doing it yourself. Such an opportunity should on no account be missed. In this case a vapour check should be laid under the insulation, but the void where the joists are can then be ventilated into the room below, as there will be no tendency to condensation provided that the insulation above covers the whole area.

A flat roof can also be insulated *on top of* the existing water-proof deck by slabs of Foamglass or non-absorbent foam plastic, laid dry and weighted down with concrete slabs or a 50 mm (2 in) layer of pebbles. There is not much experience of this technique (e.g. will the roof still drain properly, will weeds grow in it? etc.), and the original roof must be strong enough to take the large extra load.

Access hatch

If you have a loft capable of being insulated but with no access hatch, get one cut (or cut it yourself if you feel confident about the fairly simple carpentry involved). It is likely to be cheaper, or easier, to put in a hatch than to cut a hole big enough and then patch it neatly. (If you use a builder, claim not to mind about the hatch and ask him to quote for both, for comparison).

Insulate the upper surface of the hatch (lid), preferably with a foam plastic slab material, because mineral fibre frays too easily to be practical in that exposed position (unless completely covered with polythene). Put a good draught-strip (see later) around the hatch opening, and, if you can, some means of holding the hatch down to compress the seals. How you do this will depend on how you get up to the hatch: if your access ladder needs to poke up through the hatchway before it is at a safe angle, use a hatch fastener that can be released from floor level (a weight on the end of a stick, hung from a hook on the hatch comfortably within reach?).

Loft storage

Often there will be a conflict between the needs of loft insulation and loft storage. It is not advisable to simply dump stored items on top of the insulation, as at best this will disturb it and at worst cause unnecessary mineral fibre dust (see Chapter 9). You could lay boarding (e.g. second-hand floorboards, or chipboard *at least* 12 mm thick) across the joists, over a suitable area, and limit storage to that area. This will be around the hatch, normally.

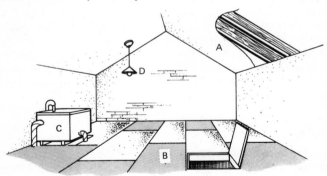

Fig. 7.7 Tent enclosure over loft storage area: foil-backed building paper (A) stapled to rafters, floor covered with chipboard (B) at least 12 mm thick. Tank(s) within enclosure still insulated (C); adequate lighting (D) controlled from below. Area should be as small as possible; take care not to overload floor – it will not take ordinary room loads

Consider forming a 'tent' around this area, consisting of polythene or (better) foil-backed building paper stapled to rafters (*Figure 7.7*). This will help protect goods from condensation, and consequent damp and rust. If the 'tent' is a complete enclosure you could omit the insulation under it, thus making sure it stays slightly warm; it should also stay fairly dry if the hatchway and other holes from below are well sealed. Such an enclosure should if possible include water tanks, so as virtually to guarantee against freezing.

Wall insulation

The examples analysed in Chapter 3 showed that walls can lose from 19 to 33 per cent of the heat, depending on the shape of the house. Wall insulation, if it is to be worth doing at all, should reduce this by a half or more. In fact if you decide on insulating any particular kind of wall, you are not likely to have much choice about the degree of improvement. There are three distinct approaches, which need to be discussed separately:
(a) cavity filling,
(b) internal lining,
(c) external insulation and cladding.
By and large this list is in ascending order of cost, although (b) and (c) can be tackled by a competent amateur while (a) cannot. Both (b) and (c) might well be done at an 'opportunity point', i.e. when the wall needs replastering internally or complete repointing, rendering or painting externally. In such a case the extra cost of the insulation is likely to be fairly modest as a proportion of the whole job. It might well therefore be cost-effective, whereas wall insulation done by itself rarely is (except for cavity fill, the cheapest method). If you insulate externally using a durable cladding, you may save future maintenance costs as well as imminent ones.

As an example, consider a room heated to a weekly average temperature of 16°C (62°F) against an average outside temperature of 7°C – a difference of 9°. A square metre of 225 mm (9 in) solid brick wall in that room would lose about 2 watts per °C, i.e. 18 watts, for seven months a year − a total of around 5000 hours; thus the heat lost through that square metre of wall would be 90 kW hours per year, of which by insulation you might save about half. (With cavity fill you would save a larger proportion of a rather smaller loss; see Appendix 5.) The saving would be worth about £1.30 a year at the price of on-peak electricity, or around 40p at the price of gas.

Remember that these figures refer to *one square metre* of such wall only. Such an estimate is probably on the high side; nevertheless, multiply it up to see whether it relates sensibly to your own heating costs, not forgetting that perhaps a quarter of your heating is provided by incidental gains, so the saving ought to be rather more than heating costs suggest. If the resulting figures are about right for you, you may choose to spend £2 per square metre (if you heat by gas) to get your money back in five years, or £4 to get it back in ten years, over and above money you would have to spend anyway for unavoidable maintenance.

Effect on thermal behaviour

You should make sure you have considered what effect, if any, a particular form of wall insulation will have on the heat storage capacity of the house, and whether or not you are likely to be able to adapt to or make use of the change (see Chapter 3).

● If you fill the cavity, the walls will warm up more quickly but cool down more slowly (as the inner skin will lose heat to the outside more slowly). They will store less heat altogether, although this will normally only affect their response to heat from sunshine through windows.

● Internal insulation will greatly shorten the warm-up time of a room, but will also shorten its cool-down time. The overall effect will depend greatly on how much *other* heat storage capacity there is in the room.

● Solid walls insulated externally will also have a shorter warm-up time because they will never again get so cold, and will lose a lot less heat to outside. Their cool-down time will increase a great deal − potentially a very useful bonus.

In all cases of wall insulation, the house will benefit rather less from sunshine falling on the outside of the walls and slowly passing through them. All the above effects are very difficult to calculate without a computer, and even then the result would probably not be accurate. But is is worth knowing roughly what to expect.

Cavity fill

Conventional cavity walls can be filled with a plastic foam called urea-formaldehyde, with various other plastic foams, with loose mineral fibre, or with beads or chips of plastic foam. The first is the cheapest, and a single house will probably cost between one and two hundred pounds for the cheapest process. Holes are bored in the outer skin of brickwork and the insulation is installed by pumping or blowing it in (blown insulants need bigger holes). Afterwards the holes are 'made good' to as near a match as possible to the existing surface. The installer should check for leakage of filling from the cavity. Warn him if you think you have first-floor joists built into the inner skin of brickwork. You should also fully understand the general risks described in Chapter 9.

Internal linings

The techniques here are rather similar to those described for flat or voidless roofs, except that normally the existing wall will be brick or concrete block. (If the wall is of timber frame, it would probably pay to strip off and replace the lining, filling the voids with mineral fibre.)

The technical problems to be solved with wall lining are:
● to accommodate the extra thickness, including enough insulation to make the job worthwhile;
● to cope with existing joinery such as skirtings, picture rails, door and window architraves and sills, etc;
● similarly to modify any wiring and piping on or at the external wall;

- to finish with a smooth and robust surface that can take knocks and can be redecorated in future;
- to avoid creating a fire hazard;
- to minimise the risk of interstitial condensation that will not dry out of its own accord before it starts rot.

The last problem is a major subject, and is dealt with in detail in Chapter 9. You should realise that although quite a lot of insulating linings are being used now, and numerous materials are available, there is not sufficient real experience over enough years to guarantee its soundness and freedom from rot over the remaining life of a house. Laboratory tests are of limited value because houses present such a range of different conditions. Appendix 5 tabulates a range of linings, with an estimate of their final U-values.

One of the cheapest and simplest linings is simply a layer of plasterboard fixed on 20 mm (1 inch planed) or 25 mm (1 inch sawn) battens. An improvement is a layer of foil in the cavity, using either foil-backed building paper or foil-backed plasterboard. In either case the foil faces across the cavity from inside. However, if condensation takes place within the plasterboard it can cause the foil to corrode and disintegrate. Also beware of the dangers of such foil in relation to electrical circuitry — your earth system must be sound and all metal boxes convincingly earthed. So if you have to 'bring forward' power points and switches, bring forward their boxes as well.

A lining will perform better if, instead of a cavity, there is a thickness of foam or mineral-fibre mat insulation. With the latter you need battens, with the former you do not, but certain disciplines are advisable to reduce the fire risk, as follows.

- A combustible plastic insulant should be covered with a fire-resisting facing.
- There should not be voids in the construction such that combustion air can reach the plastic foam (battens just for this purpose may be necessary at the top and bottom of the lining — *Figure 7.8*).
- The whole should be physically fixed to the wall behind so that, if the plastic melts with heat, it will not all fall away from the wall. This can be achieved by (say) six long screws through each sheet, into the wall behind.

It is desirable to dispense with battens if possible, as they are expensive to buy and time-consuming to fix. Where they are necessary, they are best fixed with masonry pins or screws and plugs, and it is advisable always to use timber treated against rot. You can use battens with mineral fibre stapled between them, if you can match insulation widths with plasterboard sheet widths.

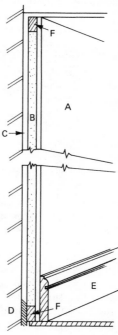

Fig. 7.8 Reducing fire risk when dry-lining: plasterboard (A), fixed to wall with six screws per sheet, backed with foam plastic insulant (B); vapour check advisable. Existing plaster (C) should first be made good at (D) after removal of skirting, which is subsequently replaced on inside (E). Continuous battens (F) are included at top and bottom of foam plastic to act as fire stops

To line without battens, the insulant itself needs to be firm enough to give adequate support to the facing material. Probably the most practicable combination (available as a proprietary product) is expanded polystyrene and plasterboard in a laminated sandwich, which can include a polythene vapour check between the two. The ideal dry-lining technique has yet to appear on the market, but this could happen at any time. It needs to be simple, robust, fireproof and cheap.

Existing joinery presents a test of ingenuity. It is desirable to remove as little as possible, but not advisable to bury any of it beneath the lining, as in such a position it is more than usually liable to rot. Hence it is better to remove, and subsequently replace, skirtings and picture rails, but work around architraves. *Figure 7.9* offers some hints at detailing around windows. Presumably you will not attempt such work yourself

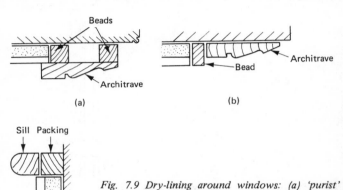

(a)   (b)

(c)

Fig. 7.9 Dry-lining around windows: (a) 'purist' approach, architrave refixed over beads to thickness of dry lining; (b) 'working around' existing joinery, architrave not moved; (c) old sill refixed over packing

unless you are a reasonably competent woodworker. It is most advisable to work out all the details on paper before even choosing the insulating material (which will determine overall thickness). You should also check the trueness of wall surfaces at this stage, too, because this can have an influence on whether a material works easily or only with great difficulty, in particular at the joints between sheets. Perfectionism is a great waster of time: if it would take enormous effort to preserve the decor of a room (fine joinery, rich plaster cornices etc.) and insulate it as well, it may be common sense to avoid insulating a particular wall altogether or parts of it (e.g. go no higher than picture rails). Take steps if necessary to cure an uneven 'radiant balance' if it becomes noticeable later.

If you propose to employ a builder to install insulating dry lining, be prepared to negotiate. One with experience of dry-lining techniques will do it well, quickly and efficiently but may not take to 'wet' plastering, and vice versa. What a particular builder is used to will be reflected in his price for the work. Also, since insulating dry lining is a fairly new process, some materials are hard to come by. Study building trade magazines for the addresses of manufacturers, and get from them the addresses of stockists. A small builder may be unwilling to travel some distance to collect materials, so having decided with the builder what you are going to use, it may well pay you to organise supply and delivery yourself. Many small builders hate organising things by telephone or letters, and are quite capable of mishandling negotiations that would be second nature to any business executive or secretary.

Finally, and above all, study Chapter 9 before going in for dry lining.

External wall insulation

Insulating the outside of a wall may be a very sensible procedure if:
● you don't mind changing its appearance entirely;
● you can get planning and Building Regulations consent (if necessary) to do so;
● the wall or walls in question are large, simple areas, preferably with an overhanging eaves at the top;
● the process does not involve too many complications such as rainwater down-pipes, windows, doors, balanced-flue terminals etc.;
● you can choose a cladding material that will not increase the risk of fire spread near a boundary (Building Regulations).

It is necessary to weigh up the pros and cons. The circumstance most likely to favour such an investment is when the particular wall needs work on it requiring scaffolding and a fairly expensive maintenance process that could be avoided by re-cladding over insulation.

The choice of materials and techniques is rather similar to that for internal lining, but of course the cladding has also to be proof against wind, rain and physical damage. Manufacturers of sheet, strip and slate-type claddings normally publish hints for keeping rain out at the edges and around door and window openings. They may also offer simple clips or brackets for battens, which need to be deeper than usual to accommodate enough insulation to make the job worthwhile. You may need a rainproof 'breather paper' laid on the battens immediately under the cladding, but on no account use something vapour-proof here as it is essential for a masonry wall to continue to release vapour outwards.

There are already some proprietary external insulating systems on the market, using various techniques, including at least two that consist of a thick render incorporating poly-styrene beads. Probably none of these can be used at a price that is likely to be recouped within ten years from fuel savings alone, but might possibly be viable in a combined insulation/ maintenance operation. Make sure you use your own estimates of likely fuel savings. Also beware of new 'wonder' materials: sun, rain, frost and wind are frequently more destructive than chemists anticipate. A variant of a well-tried process is always safer (e.g. tile, slate or timber cladding).

Silicone treatment

Merely treating brickwork with a water-repellent silicone can in some circumstances make a significant difference to the U-value of a wall, simply by keeping it drier. Provided the wall and its pointing are in good condition, the silicone will prevent or at least reduce its absorption of water, and presumably the more porous the brick the greater this effect will be. But bear in mind that it is the *average* wetness of the wall that matters, so only a wall that at present tends to stay wet for most of the winter will be greatly improved. In such conditions a reduction of 10 per cent or more in the wall's U-value might be achieved. Silicone is easy to get and quite quick and cheap to apply, but needs renewing once every few years. Once dry, it is invisible. Silicone can, however, do more harm than good to some kinds of brick — get expert local advice about your kind of brick first.

Ventilation control

Chapter 3 indicated the range of ventilation rates that occur in practice and what their consequences are in terms of heat loss. But the way such ventilation is distributed through the house is very individual to the house. Some houses have holes where others do not.

Obviously you will need to keep some air flowing, but there are very good reasons for trying to block up all the leakage that does not have a specific purpose or justification, and then providing the ventilation you need at places where you need it — above all by methods that you can control to a fine degree. Recent research has demonstrated the following.

● Air infiltration through even quite new houses tends to occur more through unexpected places than through those places most mentioned in published advice on 'draughtstripping'.

● Frequently a great deal of air leakage takes place between the house and the loft. This tends not to be noticed (you are never aware of warm air leaving a room, only of cold air entering it) and is often via cupboards, such as airing cupboards where pipes tend to be concentrated. As well as wasting heat, such leakage greatly increases the chance of roofspace condensation. Wind is likely to cause suction in the loftspace, and the buoyancy of warm air adds to this effect by pushing it out at the same places.

● Outside windspeed nearly always has a dramatic effect on ventilation rate, turning a small leakage into a large one.

● Opening windows is far too crude a method of controlling ventilation to make sense, now that energy is scarce.

● People who are out all day tend for security reasons to prefer to shut windows, leaving insufficient ventilation in calm weather.

● Gentle ventilation while the house is empty is the best way of drying out moisture that has been absorbed by plaster, furniture etc. during the periods of highest humidity, which usually occur while the house is occupied.

All the above lessons seem to support the argument for sealing all the non-controllable leakage and providing controllable 'trickle ventilation' to give a modest air flow, with no threat to security, and adjustable to suit outside windspeed. Try to pursue this principle, but bear in mind that fuel-burning appliances need air to feed them, and to fail to provide it can be dangerous or even fatal.

Gaps between ceiling and loft

The section on roof insulation earlier in the chapter dealt with the need for sealing air leakage into lofts, but did not go into methods. While the easiest materials to use may be sticky mastics, it is not advisable to apply these where they might come into contact with plastic pipes or cables. Proprietary filler from a tube is very convenient and fairly stiff. If the gaps are so big that the filling sags, use some sort of scrim (any net-like cloth will do) to hold it together until it sets. Such fillers usually set very white, so quite often they need no decoration afterwards. A loft hatch needs foam or rubber draughtstrip, but make sure you can close the hatch properly afterwards and hold it shut, without wrenching the hinges. There may be gaps at the tops of partitions in the loft-space, but usually only where the top member is drilled for an electric cable dropping to a switch. A gap-filling aerosol foam, new on the market, may be the ideal material for all ceiling gap-filling.

Gaps around windows

There are so many different sorts of window that they cannot be dealt with comprehensively here. However, a great deal has been published already on the subject, although some of it misses out on certain basic principles, such as:
● Test out the materials you intend to use, on one whole window, before buying in bulk. In particular, make sure it

works on all sides of the window without leaving gaps or straining the hinges, catches or sash/casement frame. If it does any of these things, look for a different method.

● The best time to draughtstrip windows is shortly after painting them, but not less than two or three weeks after, because the solvent in the paint may affect the draughtstrip.

● Remember that foam plastics usually have quite a short life: PVC foams can be expected to last better than polyurethane foams (the latter are the softest ones, with large cells). Sunlight destroys polyurethane quickly, and it tends not to recover after having been compressed for some time. PVC is better but not perfect. An imported seal of ribbed soft rubber is now available, with a five-year guarantee.

● The sliding faces of sash windows are especially problematic, although the amount of leakage will depend greatly on their condition. A really loose sash may be curable by nylon pile seal if there is room. Extruded rubber 'hollow bead' seals may also work. For the worst sash windows, some sashes may best be left sealed all winter with masking tape; leave one per room for ventilation, unless you can introduce an adequate ventilator into the sash.

● Louvres are frequently very draughty. Sometimes they have been fixed intentionally so that they never close tightly. If it is safe to do so, you can cure this by moving the top or bottom weatherstrip, whichever is preventing them being closed completely.

● Check and seal any gaps around the outside of frames, where they meet the wall: mastic on the outside, interior filler inside.

Gaps around doors

These behave similarly to windows, but are more likely to have inherent 'structural' problems, such as loose hinge screws, sagging frame joints, warping, and worn thresholds. Attempts to draughtstrip a bad door may just make things worse. Consider rebuilding (or having rebuilt) a bad door, and take the threshold seriously as it is almost invariably the biggest leak.

Bronze weatherstrip is the best, but often means that a lot more force is needed to open and close the door, especially at times of the year when the timber swells. You should consider this before you use this form of draughtstrip: will everyone using the door be strong enough? Even if they are, you may need to add stronger handles. Practical problems such as these tend to be ignored when considering the cost of draught prevention.

Do not forget the letter-box. In the old days people used to have roomy boxes fixed on the inside to catch the letters. This was not just to prevent the dog eating the incoming mail: it also contained the enormous draught created by a newspaper rammed half-way through and left there. A slightly less effective but cheaper method is a heavy cloth, but postmen do not like these. A heavy-gauge polythene bag with the bottom cut off may work. As with windows, check for gaps around the door frame where it abuts the wall.

Cat-flaps are virtually impossible to draughtproof. Maybe one day someone will invent a good one that closes tight and stays closed, and lasts.

Hollow floors

Timber suspended floors can admit a lot of draught, both between boards (even through a carpet) and under skirtings. Filling gaps between boards is a long, soul-destroying process. The best consolation is perhaps to use the cheapest material: a mixture of fine sawdust, plaster of Paris and size is worth trying. You will need to experiment with proportions, and also clean up as you go, to avoid having to sand the floor afterwards. The same home-made filler will do for gaps below skirtings, although sometimes they are just too big. Under a carpet, of course, masking tape will do.

You may find, if you can lift a floorboard to check the temperature of the void underneath, that it is quite surprisingly warm. This will be the case if:

● not a great volume of air is coming up into the house, to be replaced by cold air from outside, or

● not a great deal of fresh air is passing through the void, as the airbricks are not very generous and the house altogether not too exposed;

● a lot of heat is being supplied to the space, maybe because not many rooms are carpeted or otherwise insulated, and/or there are a lot of uninsulated central heating pipes down there.

You will appreciate that the more heat you are losing to the underfloor void, the more you will get back in relatively warm draughts, as well as reduced heat loss through the floor. But quite clearly the crucial source of loss is the through ventilation of the void. Therefore it is still good policy to reduce air leakage from the void upwards as much as possible, because it is all part of the general flow of air through the house, especially that caused by 'stack effect' in cold weather. You should make sure, however, for the health of the timber under the floor, that a reasonably clear flow of air through the airbricks is maintained, independent of the flow formerly going in at one side and up through the floor.

Upstairs floors

The possibility of gaps leading from a wall cavity into an inter-mediate floor void (*Figure 7.10*) was implied in the passages on cavity fill. The only way to check for draughts from this

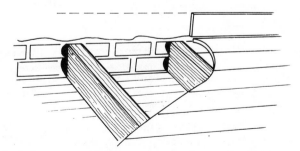

Fig. 7.10 Possible draught routes around first-floor joist ends, where joists are built into inner skin of cavity wall and gaps were not filled during building

source is by a candle flame (with care) at skirting gaps on outside walls. The condition can normally only occur where floor
joists (which always run *across* the direction of floor boards) are
built into the inner skin of a cavity wall. If you do have such
leakage, treat it as for ground floor draughts but realise this
treatment is only superficial. If severe, leakage will also affect
the heat losses through upstairs floor and downstairs ceiling.

Solid floors

Draughts at skirting level of a solid floor are a common characteristic of houses with timber-framed inner skin construction
and of some others with badly jointed brickwork below plaster
level (*Figure 7.11*). (Normally there is nothing to prevent

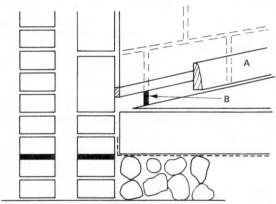

Fig. 7.11 *Draught source around solid concrete floor:
skirting (A) may shrink after building, leaving gap at
bottom and allowing air to flow through badly filled
joint (B) below level of plaster*

draughts entering the cavity — they help keep it dry.) The flow
of air was perhaps not evident when the house was built, but
the skirting will since have shrunk, leaving a gap underneath.
Remember that a 2 mm gap 4 metres long amounts to 80 square
centimetres. Treat as for other skirtings.

Power points and switches

Test also for draughts at power points and switches in external walls. If you find them, switch off all the power, remove the front plate and ease it forward far enough to enable you to tape over all visible gaps in the metal box behind. Take great care not to twist or damage cables in the process. If you are not used to electrical work get an electrician to do this for you. A foam filler may do this job well.

Flues

If you have old flues that are not in use, block them off with hardboard or chipboard. Advice is usually given to incorporate a ventilator 'to keep the flue free from condensation', but that is *not* the best thing to do unless the ventilator is an airbrick low down on the *outside*. A cold flue is more likely to feed air *into* the room if it can, and will not cause condensation unless the downdraught is of damp air and the flue extremely cold, which will occur only rarely for short periods. If your ventilator is high up in the room on the inside, a stack effect up the flue may develop, in which case the ventilation will cause rather than relieve flue damp: warm, moist air will be taken up the flue, to condense on its colder surfaces.

If you break into a flue to fit an airbrick, either in the room or outside, you may write off the flue for future use, unless you place the airbrick low down behind a (removed) firebrick. A sound flue is a very valuable asset, not to be lightly abandoned for good. Who knows when you or someone else may need it in future?

Cellar doors, hatches etc.

Treat as appropriate. In some instances there is a case for some *ad hoc* insulation work — for example where an indoor solid-fuel store has a rather flimsy outside door or hatch. Improving these should be taken seriously, as they can lose a lot of heat, but make sure you do not create an 'unventilated cellar.'

Interior draughts

Most advice on draught control assumes whole-house heating, where only the draughts to outdoors usually matter. But for many people, the draughts between cold and warm rooms are just as important. Here you have to try and envisage what the internal airflows are in the house.

You can indeed save a lot of energy by containing warmth in a few rooms. You can also improve the comfort of rooms by eliminating draughts. But bear in mind that any rooms that stay stone cold are more likely to suffer from condensation, which ventilation may not help because it may simply cause warm, moist air to pour into the cold rooms and deposit more damp. If you do genuinely want to let some rooms go cold, provide them with a little ventilation to the outside, but at the same time take steps to seal their doors from the rest of the house to prevent such a flow of moist air. However, rooms totally without heat (e.g. by transfer through walls or floors from heated rooms) are likely to become damp anyway.

As to draughts in warmed rooms, see Chapter 5 on the nature of comfort. Remember that it is the feet that suffer first, and the cold draught under a door is the worst. The cloth-covered 'sausage' draught excluder is now unfashionable, but it still works well and is quite easy to make.

Trickle ventilation

The principle of trickle ventilation has been discussed elsewhere. The question is how to achieve it and how much to provide.

Slim ventilators that provide a suitably small airflow are now made for fixing into the opening sashes of good-quality windows. Some consist of a long, narrow aluminium 'hit-and-miss' inside part, for mounting over a line of holes bored through the timber (*Figure 7.12*). The outside part, which is essential, consists of a plastic cowl incorporating a mesh insect screen. These ventilators are not expensive and are quite easy to fit to some windows or to external doors. Other types of window would need modification, and hence some ingenuity on your part.

As a rough guide, you will need around 15 square centimetres (2½ square inches) of ventilator 'free area' per 10 square metres (100 square feet) of room area (sheltered sites ought to have more, exposed ones less). This area must be progressively closable down to zero. When more ventilation is wanted you can

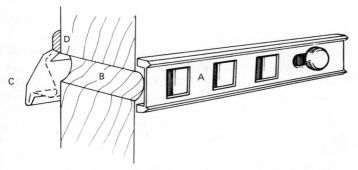

Fig. 7.12 Proprietary 'hit-and-miss' type ventilator suitable for trickle ventilation: inside component (A) with adjustable hit-and-miss operation, is fitted over row of holes (B) drilled in window sash or frame, or in door. Outside cowl (C) with mastic channel (D), has insect screen

then open a window a crack as well — but get out of the habit of trying to freshen up a room within a few minutes and then leaving open that much ventilation for a long period. Trickle ventilators ought to keep rooms reasonably fresh all the time with the minimum of heat loss. They will need closing when it is windy or when the room internal door is left open.

Such ventilators may not be easy to come by at present, but should soon become more readily available (see Appendix 4).

8 Insulation (2)

Windows

You may recall, from *Table 3.1*, that the proportion of heat lost via the windows of the room shown in *Figure 3.3* was estimated at 28 per cent. It is not difficult to imagine a similar room with big patio doors where heat loss would approach 35 per cent of the total or even more. Such figures are not uncommon for living rooms with big windows, where the window to wall ratio can be 40—50 per cent, whereas the average of this ratio over a whole house is more likely to be 10—15 per cent. The rooms with the biggest windows also tend to be those heated to the highest temperatures.

For these reasons the case for double-glazing living-room windows is often very strong. It is especially so where:
● the living rooms are heated to a high temperature all day, when curtains are not pulled;
● air temperatures in rooms with large windows are kept extra high to counteract the 'negative radiation' of the large, cold glass surface (see Chapter 3);
● the heating system for the whole house is controlled by one thermostat, which is set high enough to suit the living room.

Clearly, therefore, any analysis of the economics of double-glazing based on whole-house temperatures, heat losses and costs is liable to be grossly misleading. Consider your own living room(s) and decide on that basis.

There may not be a sound economic case for double-glazing in other rooms, but it is a useful possibility to be examined for improving the comfort of individual rooms for specific purposes. Always consider insulating shutters (see later) as an alternative: they may make more sense.

Bay windows

The large bay window is a feature of perhaps millions of older houses. The overall glass area may or may not be very great, but the small areas of wall between windows will tend to have greater heat losses than similar areas of flat wall, so the total heat loss from the whole bay is quite considerable.

Such bays lend themselves to DIY double-glazing, because the individual windows are usually of a handleable size compared with the enormous 'patio doors' of more modern houses, for which there is no real alternative to a professional job. Furthermore, if the bays have vertical sliding sashes you can double-glaze some of the windows, leaving others (typically one or two out of five) openable for occasional ventilation during the winter – unless you can fit trickle ventilation that will allow you to double-glaze the lot.

The piers between the windows and the strips of wall above and below still lose a great deal of heat, so it is worth considering installing insulating shutters that fill all or most of the bay. These are discussed in detail later.

Performance of double-glazing

Quoted figures for the performance of double-glazing vary quite a lot, depending largely on whether both window and frame have been taken into account, and what degree of exposure (to cold winds) is assumed. Also, a window with a low horizon in front of it will lose more heat to a clear night sky. With all these factors to consider, quite clearly generalisations are bound to be very rough. With those caveats, typical figures could be:

Single-glazing $U = 4.3 \text{ W/m}^2 \,^\circ\text{C}$
Double-glazing $U = 2.8$
Triple-glazing $U = 2.0$

Triple-glazing appears to be less likely as a good investment, but it is included because if you are doing a DIY job in clear plastic with frames, the extra cost is quite small and the whole can be achieved for less than the cost of double-glazing in glass.

Professional double-glazing systems

With professional systems you now have a wide range of choice, with little or no difference in performance. Such differences as there are include:

● Different spacings between the sheets of glass in 'sealed units': a space of 12 mm (½ inch) is materially better than 8 mm, but above 12 mm the improvement with thickness rapidly diminishes towards nil (although big gaps perform better for sound insulation).

● The incorporation of insulation in the frames and sashes: obviously this is an improvement, but what proportion compared with an all-metal or all-wood window? Is it worth the extra cost?

● Liability to condensation, either between the panes, or on the frames and sashes: sealed-unit double-glazing should very seldom have any on the glass, except on the *inside* (room) surface when there is extreme inside humidity. Other systems nearly always have some on occasions. It matters only if constant or damaging.

● Varying ease of cleaning, maintenance and ventilation.

If you are buying replacement windows, double-glazing is probably well worth it, especially if you will end up with less maintenance to do. There may be a tax advantage in these circumstances; on the other hand, you may find that your rateable value goes up (if the valuer notices), indicating that valuers find double-glazing adds to the value of the house.

The range of systems can be listed as:

(a) Double-glazed replacement windows.

(b) Secondary sashes: added sash by sash on the inside, and most commonly vertical or horizontal sliding, in aluminium or plastic.

(c) Secondary windows: separate additional windows complete with frames, fitted inside and opening inwards, continental style. As often as not the frames are timber; it is worth checking the cost for comparison.

(d) Secondary sashes added on the *outside:* fairly rare.

(e) Sealed units in place of existing panes, possibly in 'stepped' form if the present rebate is not deep enough (*Figure 8.1*).

(f) 'In-situ sealed units': panes added inside the present ones without disturbing them, using a special sealing bead, with the void then dried out (*Figure 8.2*). This is a new idea from a firm based in Denmark (see Appendix 4).

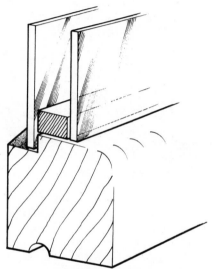

Fig. 8.1 Stepped sealed-unit double-glazing (used in place of existing glass)

Of this list, type (a) is probably both the soundest and the most expensive. Type (b) is commonest, and the price varies according to how much of the job is left to you. Some manufacturers make a great fuss about the airtightness of the frame they mount their windows in, but their sliding sashes rely on brush-contact seals, which cannot possibly be airtight. Some of the vertical sliders have no counterbalance weights or springs, and could be dangerous for old people and children. There is almost bound to be some condensation on the outer pane at times.

Type (c) should have a performance advantage from the extra frames, but if these are timber they are likely to be awkward to repaint. Also, as the gap between original and new panes is probably at least 50 mm (2 in), or maybe a lot more, the performance advantage will then be reduced by heat loss through convection currents within this wider void.

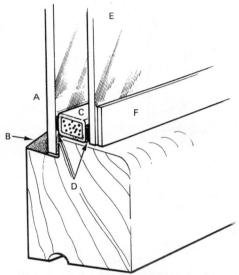

Fig. 8.2 In situ sealed-unit double-glazing: existing glass (A) is left in position, with existing putty or bead (B); spacer (C) and airtight adhesive seals (D) are inserted, then new pane (E) is added, with optional trim (F) on inside. Spacer incorporates desiccant to remove moisture from cavity

Type (d) seems always an illogical approach to double-glazing, as the components need to be detailed to resist wind and weather just as effectively as the existing window they cover up. Allowance for window opening and for effective decoration of the existing window is likely to be a problem in practice. If this technique were chosen as a way of draught-proofing, the gaps that formerly caused draughts would be on the inside and would lead to severe condensation on the outer pane.

Installing type (e) is fairly sound practice, although consideration should be given to whether existing hinges can take the load of double the weight of glass. Most sealed units are guaranteed for about ten years. They are excessively expensive and inconvenient windows to have broken, of course.

There seems to be no experience of type (f) in Britain yet. It seems a very good idea, and theoretically quite economical. It might be applied to existing vertical-sliding sash windows, with very little change in their appearance, provided that cords and weights could be modified for the extra load.

DIY double-glazing systems

There seem to be six basic alternatives for DIY application:
● Some kits allow you to self-build vertical or horizontal sliders similar to the professional type. You have to absorb the cost of purpose-making the units to suit your window. There is of course some waste of materials, and you acquire the glass yourself at retail price (about £6.50 per m²). Aluminium and rigid plastic are both available, in some quite sound designs. Plenty of time is needed for the job.
● Far simpler kits consist of a soft plastic edging strip to be fitted around the glass and fixed back with clips (or some with rudimentary hinges). The result is adequate as double-glazing, but is prone to vapour leakage, and to condensation that is not easy to mop up, and can thus set up mould or even rot in existing timber frames or sashes.
● You could install complete secondary windows inside the present ones, either purpose-made or, if you are lucky, standard sized. This is the same job as the 'professional' equivalent, and is suited to houses where the existing window is towards the outside of the wall. You lose the wide inside sill, of course.
● A DIY job with beads, using your existing windows, is possible with some windows and not others. Especially easy where old sash windows have been replaced by fixed glazing with louvres over. Vent holes to the outside (at the bottom) are recommended to reduce condensation, which you can never get at.

138

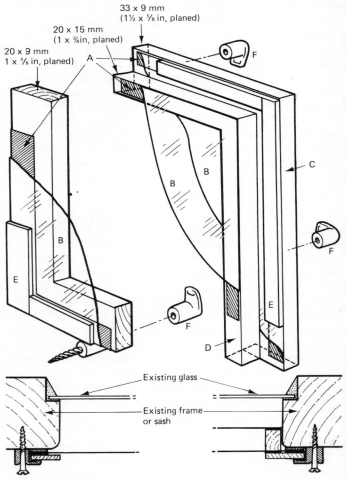

Fig. 8.3 DIY double- and triple-glazing in 'glass-clear PVC': frames are made up, fixed square, prepared with double-sided adhesive tape (A) and dropped on to the PVC (B), which is laid on a smooth, flat surface. For triple-glazing, inner unit (C) is made and glazed, and added to outer unit (D), which is then glazed. Fix complete units to existing frame or sash using proprietary draughtstrip (E) and nylon double-glazing clips (F)

● You could use 'glass-clear PVC' (*Figure 8.3*), the thickest of which costs about a quarter the price of glass. It is highly transparent but very shiny, so any wrinkles are obtrusive. Best fitted to a light timber frame with double-sided adhesive tape. The frame is then bedded on a good draughtstrip and fixed back with double-glazing clips. It is quite easy to fix two sheets of PVC to one frame for triple-glazing, and still cheaper than one sheet of glass. Caution: PVC is flammable.

● Even cheaper, but short-lived, is polythene fixed up with adhesive tape. This is a good emergency measure at very low cost, but rather unsightly.

Condensation and double-glazing

The UK climate being what it is, it is quite unusual to have completely condensation-free double-glazing. The crucial factor is the tightness of the seals around the inner glass. In practice there is no problem, provided that it is quite easy to open up the inner sashes and remove the water from time to time. You should avoid allowing water to lie around at the sill for long periods, as this encourages mould − the commonest being 'black spot'. A fairly weak solution of bleach will take care of mould spots for a while, but is not permanent. Do not allow water to soak through paintwork into sills: in due course they will rot.

Double-glazing will produce less condensation on the inside of the glass than single-glazing in the same circumstances. However, some double-glazing with metal frames suffers severe condensation on the frames, especially if they are of wide section. Also, in so far as double-glazing reduces window pane condensation it allows slightly higher humidity, and hence greater risk of condensation elsewhere.

Insulating shutters

There are some insulating shutters or blinds on the market at present, but only one seems to make a sufficiently great improvement to window heat loss to make it worthwhile. The best bet is still to make your own − for example, using a rigid foam

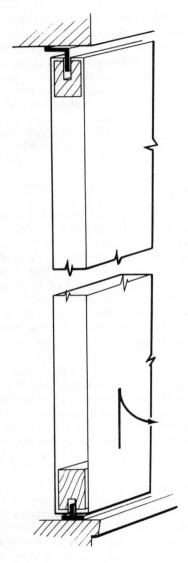

Fig. 8.4 Lift-out detail for insulating shutters

plastic board, accurately fitted and glued into a lightweight timber frame, and covered with cloth. On a small window these can be hinged, and folded back against the wall during the day. For a larger window you may be able to use hinges, or make lift-in panels located by a timber channel (or PVC or aluminium track) at top and bottom (*Figure 8.4*).

There will of course be problems in relation to curtains, depending on whether or not you are prepared to dispense with these while the shutters are in use (the shutters could of course be covered with a curtain fabric, glued or stapled in position). There will be different habits to be got used to. But such panels 50 mm (2 in) thick will improve the U-value of a single glazed window from 4.3 down to around 0.54 W/m² °C. This represents an eightfold reduction in heat loss, over the window area, while the shutters are closed. Note the remarks in Chapter 9 on the fire risks of plastic foams. You could cover shutters on one or both sides with a fireproof building board, but of couse they would then be heavier.

Curtains and pelmets

The very poor thermal performance of single-glazed windows is greatly improved by curtains, possibly to a similar extent to double-glazing. However, this improvement is partly conditional on whether the curtains fit closely at the top, sides and bottom, and to some extent also on their thickness. Note that curtains also protect the room against the 'negative radiation' (see Chapter 3) of the window; in this respect their fit is a lot less important, and their thickness rather less so.

The best curtains are those that fit within rather than over the window opening, and have a small pelmet at the top to prevent a downdraught behind them. They should also stop at a broad sill at the bottom, with no folds of cloth overhanging this sill. In practice there are probably few curtains quite like this — largely because they look rather mean. A box pelmet is perhaps the most important feature in other circumstances, as a positive means of preventing a downdraught behind the curtains. The ends of the curtains should be as close to the wall as possible, preferably touching it, or alternatively stopped at an edge batten (*Figure 8.5*). The bottom edge should just touch the floor or sill, to prevent the pleats of cloth providing tubes down which cold air can pour.

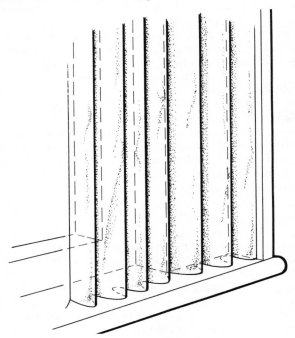

Fig. 8.5 Curtain end-stop on extended sill

There is another possibility for insulating windows that is worth mentioning just for interest. It is a wide double-glazed window, the cavity of which is filled with expanded polystyrene beads during the night, and cleared during the day. This is done mechanically using an air pump and a large storage container.

See Chapter 9 for fire-escape and fire-risk considerations. Remember that PVC double-glazing burns readily: think twice about adding a fire risk to any that might already be present (e.g. in the form of unproofed curtains). If you are sure that fire is not a risk, then PVC double-glazing may be a good choice for a small child's room, being a great deal less easy to break. But children grow up, and most go through a phase when they play with fire. PVC double-glazing is theoretically less durable than glass, but there is no reason why it should not last indefinitely, given that the vapour seals are renewed once every few years, or are made of a 'permanent' material, such as soft rubber.

Doors

A solid timber external door 40 mm (1¾ inch nominal) thick has a U-value of around 2.1 W/m^2 °C, similar to that of a 225 mm (9 in) solid brick wall. If the door is glazed its U-value will be a lot higher, representing a much greater heat loss. A door with panels will similarly lose a lot more heat, even if you can not see through it. Reducing this loss is a problem, especially if the door is decorative and you are loth to change its appearance. Hence there is a powerful argument for incorporating a 'draught lobby' of some sort, and this possibility is worth thinking over. Such a feature, added on the outside or the inside (whichever is more convenient), can make a far greater impact on heat loss if it forms an effective air-lock between inside and outside. For this purpose there must be enough space between the two doors for one to be closed before the other is opened, and room between the two for you to put down what you are carrying to unlock the inner door. The letter-box should be in the outer door, otherwise someone delivering mail might leave it open out of annoyance.

Floors

The heat losses from floors were discussed in Chapter 3. There are 'conventional' U-values for suspended and solid floors (0.2–1.1 and 0.2–1.5 W/m^2 °C) according to their size and the amount of external wall perimeter (these values are used in conjunction with inside and outside temperatures). Heat loss in practice also depends on the following:

• for suspended floors, the ventilation rate of the space underneath, which can vary greatly;

• for solid floors, the average temperature of the concrete slab, which will be much affected by how continuously the room is heated.

If the loss of heat through the floor is large, you will almost certainly be aware of it in terms of comfort. Make sure, however, that it is a problem of conducted heat loss and not of cold draughts, and act accordingly.

If the floor is very cold, adding insulation on top will increase the risk of condensation at the cold surface underneath. In extreme circumstances this can even occur under a carpet. Condensation is most unlikely on a suspended floor, because the timber boards themselves will never get too cold on the surface. The insulation will have to consist of an insulating material and a harder top surface strong enough to spread the load of concentrated weights such as the feet of furniture or high-heeled shoes. The softer the insulation, the stronger the top sheet needs to be. Experiment before deciding what to use.

Having made the decision, if the floor is concrete you can test for the likelihood of surface condensation by leaving a fairly large sheet of the insulating material flat on the floor for a few days (in winter). Place it at the edge of the floor along an outside wall. If when you lift the sheet the floor or the underside of the insulation is damp, then you will need a vapour-check membrane *above* the insulation, or an insulating board that serves this purpose (e.g. bitumen-impregnated fibre insulating board if you can get it). Polythene sheet is the obvious material to use as a vapour check. Either lap all the joints at least 200 mm (8 in) or seal them with tape. There is probably no other realistic way of dealing with the outer edges than tucking the polythene down the edge of the insulation and sealing the crack with a sticky mastic (*Figure 8.6*). The mastic will then be covered by the surface board.

The best surface board is hardboard. You cannot fix it down, but as long as there are no small pieces and it all fits quite tightly, it will work. You should, however, leave a gap about the thickness of a coin at all edges to allow for expansion when the room is humid. For sufficient strength you will need two layers of 3.2 mm ($^1/_8$ in) hardboard, with the joints staggered, or one of 6.4 mm (¼ in); the former is stronger. It is inadvisable to lay on top of this a tiled finish that needs to be glued down: where the hardboard moves with humidity the joints will show.

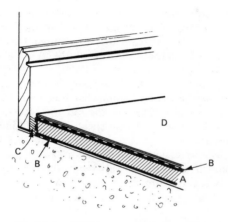

Fig. 8.6 Insulating concrete floor: 12.7 mm (½ in) insulation board (A) is covered with polythene vapour-check membrane (B), which is tucked under and sealed with mastic (C) at the perimeter; two layers of 2.5 mm ($^1/_8$ in) hardboard are then laid over the top (D)

Sheet materials or carpet are better. If you lay such a floor in a kitchen, bathroom or w.c., great attention must be paid to sealing around the edges of the finish, otherwise spilt water will penetrate down to the vapour barrier and the chances are that the wet hardboard will begin to smell. Just remember not to fix down anything that will puncture the polythene. If you have decided that a vapour-check membrane is not necessary, you have far more freedom as to what to put on top.

It will be necessary to cut a strip from the bottom of any doors opening into the room, so that they will clear the now higher floor, and also to insert a threshold to protect the edge of the insulation, surface board and finish.

Radiator reflectors and shelves

Loss of heat from the wall immediately behind a radiator will be somewhat greater (by perhaps half) than the loss from the rest of that wall. However, in the worst possible case − on an outside wall − the extra heat loss is not likely to be more than about 25−50 watts from a large (1½ m² or 15 ft²) radiator, i.e. about 1−2 per cent of the heat losses from a '2½ kW' room.

This loss can be reduced somewhat (say by half) by fitting a sheet of reflective foil behind the radiator. It will work best if the foil is mounted on a thin sheet of board (such as thick card) fixed flat against the wall behind the radiator. The whole operation will be easier if the radiator can be removed from the wall, and if you can do this then take the trouble to insulate the brackets from the wall (e.g. with pads of cork) as well, since these conduct heat from the radiator direct to the wall.

Before installing a radiator reflector, however, consider whether it is important that the wall behind the radiator, besides absorbing heat, also stores it and re-radiates it later. If you rely on the thermal mass of the structure to smooth out the ups and downs of radiator temperature, it may be sensible not to waste time with this improvement at all.

The column of warm air rising above a radiator will create a pool of warm air at the ceiling. A radiator shelf throws this air away from the wall and reduces the 'hot pool' effect, giving you a more even air temperature. Test for the 'hot pool' phenomenon with thermometers at various heights (*Figure 3.10*). If there is more than 2−3°C (4−6°F) difference between floor and ceiling, radiator shelves are probably worthwhile. But remember that the effect will slightly reduce the supply of heat through the floor to the room above.

Party-wall insulation

With or without thermometers you will be well aware of it if a party wall is cold, signifying that your next-door neighbour's house is significantly colder than yours. Very few party walls have a U-value better than about 1.8 W/m² °C. Whether or not their coldness is perceptible at normal times, it will certainly be so if the neighbour's house is left unheated for a few days in winter, and you may then need extra heating to be comfortable. So there may be good reason for you to consider insulating all or part of the party wall.

In inner suburban houses with a narrow plan, the party wall or walls may be very long, constituting a large proportion of the heat-storage capacity of the house. If you want to reduce this, insulate it. But remember you cannot insulate it without a consequential reduction in its thermal capacity, which you may or may not want.

Condensation is most unlikely to be a problem, so a simple, cheap treatment is possible. Don't use very thin layers of polystyrene, but 12.7 mm fibre insulating board will probably serve well, and should reduce the U-value of the wall from about 1.8 down to about 1.25 W/m² °C. Try to buy the denser, smoother version of this board, even if it insulates rather less well than the lighter, softer type. The former will take a paint finish better, can have its edge bevelled to make for neat joints, and neat edges at skirtings and picture-rails, and is far less easily dented. Fix it to the wall with adhesive, if the wall is flat enough. Otherwise use screws.

Curtaining the staircase

It was explained in Chapter 3 how the principal effects of roof and wall insulation upstairs in a 'partially heated' house are to make rooms on that floor warmer, but with very little energy saving. The full savings can only be achieved if the temperatures of the upstairs rooms can be kept down to what they were before.

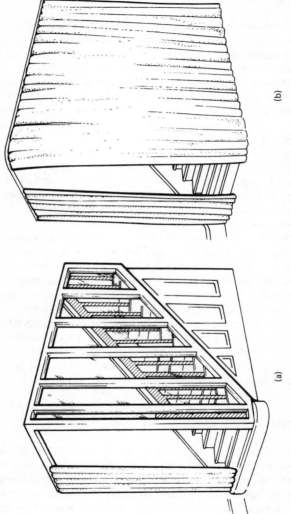

148

(a)   (b)

Fig. 8.7 Curtaining-off staircase: (a) glazed side screen with curtain at foot (provision must be made for later redecoration of staircase joinery); (b) complete curtain enclosure, preferably using light-coloured, translucent fabric

Some of the heat passing up to bedrooms gets there by conduction through their floor, but normally rather more is transmitted by convection up the staircase. This, then, is your most effective means of control. How easy it is to do this will depend on the design of the staircase, but for many types it will not be difficult. The openable part could be a door or a curtain: the latter is probably preferable, besides being cheaper, and could be made to look quite attractive. *Figure 8.7* suggests possible alternatives.

Government grants

At the time of writing it is possible for a private householder in Britain to get a grant of up to £50 towards the cost of hot-water cylinder lagging, roof insulation and insulation of pipes etc. in the roof, up to a maximum of 66 per cent of the total cost. You get this from your local authority, to whom you must apply for all the details.

This scheme (under the Homes Insulation Act, 1978) applies also to any private tenant, with the landlord's consent. At present it applies only to houses with no insulation at all in the roof. It is possible that in due course the scheme will be extended to cover 'topping-up' of inadequate existing insulation, and perhaps of other kinds of insulation as well, although this may take some years to arrive. One of the stumbling-blocks is that the Government is to some extent legally responsible for the consequences of anything it grant-aids in this way, and there are technical hazards associated even with simple roof insulation, as Chapter 9 shows; this applies even more to many other forms of insulation.

Houses owned by local authorities are being improved in similar ways — in fact a rather greater proportion than in the private sector, because the work is largely being done under contract, so the tenant has no work to do. (Private householders may employ contractors also, but then the grant will not stretch so far.) Some council houses are being draughtstripped also: this was not practicable under the private-sector scheme. Old or disabled people and those of limited means also have special status here: again, ask the local authority.

Sources of materials

Appendix 4 is a list of some of the components and materials that are mentioned in various chapters, and are not yet readily available. No attempt has been made to provide a comprehensive list of products or manufacturers. New products for energy saving are coming on to the market almost daily, so any such list would inevitably become out of date very soon.

If you have to contact a manufacturer direct, ask him for a list of suppliers local to you. It will normally be preferable to buy locally than to deal direct, unless there is no alternative.

9 Structural and Other Precautions

A house is a very complicated thing, if you take into account its structural strength, the chemical and organic nature of its materials, and the action of its mechanical and electrical services. All these are constantly being influenced by the behaviour of its inhabitants, whether consciously or not. The other all-pervasive influence is the weather. Abandon the notion that a house keeps out the weather. In fact it only keeps out the rain and a proportion of the wind and sunshine. Temperature, humidity, wind speed and sunlight constantly invade the whole building, which must be designed and managed accordingly. It is also relevant that the fabric and the services tend to deteriorate or wear out, in time, and need maintenance or replacement.

Wherever the house, its services and the household interact with each other, any change on the part of one of the three is likely to affect the others. Hence if you change the house or the heating system or your habits in order to save energy there are likely to be side-effects, because other parts of the whole system will start to exist in different conditions. Most of these changes will not matter, but some will. So unless you become rather more aware of what is happening overall, you could find yourself in trouble.

Many of the hazards associated with energy-saving improvements are already understood; warnings as to the necessary precautions are commonly disseminated. An example of this is the advice to lag the water tank and pipes when insulating a loft, because the loft will become colder and the chance of freezing will increase. Other hazards are less well known as yet, because evidence is less easy to get at, not much experience is available, and some problems take some time (perhaps years) to come to light.

When you insulate your house you are adapting it to suit a changing world. But the responsibility to be vigilant for possible consequences inevitably lies with you. The professions and the Government will of course do their utmost to keep you informed, within the limits of their resources.

This chapter serves both to reiterate warnings and cautions given elsewhere in this book, and to add a few others, especially concerning problems about which not very much is known at present.

Condensation and damp

This subject has been covered quite thoroughly elsewhere, because it is a problem that arises throughout the structure of houses, particularly in the moist climate of the British Isles. Where it occurs within a room it is easy to keep an eye on, and is unlikely to prove very serious even in the long term. But it can also occur within the structure, where it is not noticeable, and can cause timber to start to rot, which in due course could be catastrophic. Similarly it can cause corrosion in steel fastenings, sufficient to lead to their failure after a few years.

Condensation occurs when humid air reaches its 'dew point' at a cold surface (see *Table 2.2*). Slight differences in the temperature of that surface or the humidity of the air will make all the difference between some degree of condensation and none. Different households in identical houses may behave in subtly different ways such that in one house there will be condensation and in another there will not. Better insulation may either improve matters or make them worse — perhaps affecting different parts of the house in different ways. There are no general rules yet available that will guarantee immunity, and possibly there never will be. But an understanding of what causes condensation will help you to keep an eye open for trouble.

The parts of the house most at risk are structural timbers: the roof structure, the upper-floor joists, and suspended ground-floor timbers. For roofs and ground floors the best defence is ventilation (see Chapter 7) because, although this treatment may mean more moisture is brought in on occasions, for most of the year its tendency will be to dry things out. The tendency of ventilation to keep these timbers cool will also help, because rot, which is a fungus, needs warmth to grow.

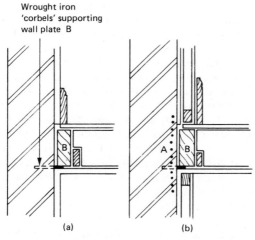

Wrought iron
'corbels' supporting
wall plate B

(a)   (b)

Fig. 9.1 Condensation risk at first-floor/wall junction. (a) Before insulation, room heat is likely to keep whole wall at a temperature above dew point. (b) After insulation, whole wall is colder, producing condensation risk within floorspace (A). Timber wallplate (B) is at risk of rot, being in contact with brickwork

Condensation does not usually occur in intermediate floor spaces, or rot in the timbers here would be more common. However, insulating walls by internal lining could in theory cause trouble, because the inner face of the wall in the floor space will then certainly be colder. Trouble would be a lot less likely if the house were well and continuously heated. The wall surface in question is the unplastered strip (*Figure 9.1*) between ceiling and floor — the depth of the joists. This wall is slightly heated by the relatively warm air in the floor space, which is likely to stay at a temperature between those of the rooms below and above. (It will be higher where there is a pool of warmer air at the ceiling.) But the relative coldness of the walls above and below after insulation will keep it a lot colder than it was before insulation.

Take, for example, the wall surface temperature of such a wall (225 mm or 9 inch solid brick) in the author's own house: with general room temperatures around 16–18°C (61–65°F), the wall surface within the floor space is around 10°C (50°F) at an outside temperature of 3°C (38°F). This would probably drop to a minimum of 8°C (47°F) if the outside temperature were zero for several hours. It might thus be expected to drop to 6°C (43°F) or lower if the walls above and below had been well insulated. Normal house humidities are in the range 40–70 per cent at 'normal' room temperatures, giving a range of dew points of 4–12°C (40–54°F). Thus this piece of wall would be at risk in cold weather in a fairly humid house, but not in a fairly dry house or in mild weather. If there were condensation it could cause damage:

● if the wall concerned were in contact with timber not treated with preservative, and
● if the brickwork were not otherwise dry enough to absorb the moisture and cause it to migrate outwards, and
● if the periods during which condensation occurred were continuous or frequent during any one winter.

Studies of condensation in theory and in practice have tended to show that in comparable circumstances rot occurs in practice less frequently than theoretical analysis would suggest. This is probably because condensation usually occurs intermittently between drier periods, which are adequate to keep timber safe; it may also in part be due to the capacity of the wall itself to take the moisture away. But some winters are a great deal worse than others, and one particularly bad one may be sufficient to wreak havoc.

A wise policy for any householder is to guess where the condensation risk areas are, and make sure you can check them a couple of times a year (say September and May) via one or two removable floorboards. It is usually possible to tell whether timber or brickwork actually feels damp (and if rot has started it will smell musty), but there is no substitute for a professional moisture meter (*Figure 9.2*) borrowed from an architect's or surveyor's office. These normally contain instructions on their use and interpretation.

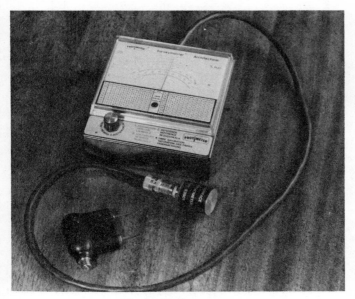

Fig. 9.2 Moisture meter

If you find evidence of condensation, don't panic. You have up to three courses of action:
● make further checks at monthly intervals to see if the dampness gets worse or better; and/or
● change whole-house ventilation habits a little, aiming at reducing humidity (you will need a hygrometer for this); and/or

● improve the ventilation of the 'risk' areas, by boring holes in the floorboards along the affected wall and also at the opposite side of the room at the far end of the same spaces between joists (*Figure 9.3*).

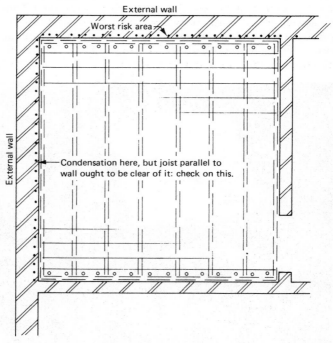

Fig. 9.3 Pattern of ventilation holes in floor to relieve condensation; cover holes with gauze to keep out spiders, and stain gauze to match floor

You should be especially cautious about adding an insulating lining to a wall that seems to be exceptionally cold, or damp, or exposed. Consider external insulation instead, or forget the whole idea. In more average conditions you are most unlikely to bring on an attack of damp so bad that it can only be cured by the desperate last resort of removing some or all of the lining.

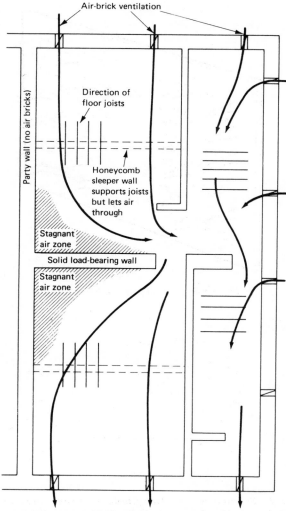

Air-brick ventilation

Party wall (no air bricks)

Direction of floor joists

Honeycomb sleeper wall supports joists but lets air through

Stagnant air zone

Solid load-bearing wall

Stagnant air zone

Fig. 9.4 Plan of walls below ground floor. Arrangement of walls within underfloor space can create zones with little air movement; these are most at risk from condensation

158

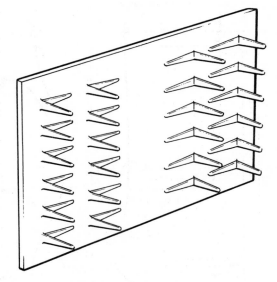

Fig. 9.5 Pressed nail-plate timber connector

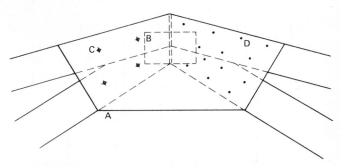

Fig. 9.6 Substituting for a failing nail-plate timber connector (B): gusset plates (A) of 12.7 mm (½ in) exterior-grade plywood, fixed either with galvanised bolts and washers (C) or bright zinc-plated screws in drilled holes (D). Always use an engineer-designed gusset if possible

In roofs and ground-floor voids the risks of condensation are low, as long as the spaces are reasonably well ventilated. But note that sealing off ground-floor draughts may reduce the ground underfloor ventilation a lot. Again, try to create inspection panels at 'damp risk' areas (see *Figure 9.4*), and if you suspect trouble you may need to provide two or three more air-bricks to improve the ventilation rate. Replacing inspection panels with well-ventilated boards for the summer will help dry out the effects of condensation over a 'bad' winter.

Metal structural fastenings

Any large increase in the dampness of timber may accelerate the corrosion of its metal fastenings. This process could also be aggravated by the corrosive action of chemicals used in timber preservative treatment. The fastenings most at risk seem to be those made of thin metal, in particular the pressed-out nail-plates commonly used on trussed rafters (*Figure 9.5*). Some of these have their spikes pressed through after galvanising, so the zinc coating is missing down the edges of each spike.

Take a close look at such plates from time to time to make sure that roofspace condensation has not caused them to rust. Radiant frosts may lead to condensation on the exposed metal, even high up in the roofspace. If the plates corrode badly it may be necessary to fix timber substitutes over them (*Figure 9.6*) before they weaken sufficiently to fail. Treat sound but endangered ones with a wax-type anti-rust spray. The fastenings in old roofs, even though not galvanised, are usually of thicker metal and more thoroughly buried in timber for protection. Some chemical fire retardants (e.g. for cellulose-fibre insulation) may also be corrosive.

Freezing

If you improve insulation anywhere within the extreme outer surfaces of the house, everything outside the insulation you have added will get colder. This change will affect:
● all water pipes and tanks outside the insulation;
● possibly the outer surfaces of walls and roofs.

The treatment of pipes in roofspaces is covered in Chapter 7, as this is part of the routine of roof insulation. Don't ignore any part of this advice. For example, failure to lag the hot water expansion pipe can lead to a plug of ice forming in it, so that when hot water is drawn abruptly from downstairs there is no air through this pipe to relieve the resultant pressure drop in the system. Thus the cylinder can 'implode' (collapse inwards).

Colder outer surfaces of walls and roofs may have no effect except in the very long term. Pointing mortar, and some bricks, tiles and slates, can be slowly destroyed by successive winters of frost attacking the porous outer layers. Some materials suffer more than others, such that they may have been suitable hitherto for use in the south but not further north. Increased insulation may hasten their deterioration. If you have bricks or tiles that seem to be vulnerable to frost, either think twice about insulating the walls (internally or with cavity fill), or protect them with a coat or two of silicone — but see the caution on pages 122–123. The same applies to tiles or slates, but you may simply have to face replacing them sooner than if you had not insulated.

Ceiling and roof loads

Most forms of insulation are so light that they will have no appreciable effect on the loading of the ceiling joists. If in doubt (long spans, ceilings already sagging or cracking off, or extra heavy insulation intended) get professional advice before proceeding.

If you are insulating the underside of a ceiling with something heavy (such as plasterboard) you should certainly get advice about whether the ceiling joists will be strong enough not to sag — and indeed advice about how to fix up the material. If a builder is doing such work for you, make sure you ask him (in writing) to consider this aspect, so that if it collapses he cannot blame your choice of materials.

If a ceiling is in bad shape already, and you seldom if ever walk around in the loft, the business of moving about on the joists to lay insulation may do some damage. If in doubt, get someone to watch the underside of the ceiling as you first walk across it, to see how much it moves and to what effect. It may be advisable to use a carefully thought-out system of planks to apply your weight near the ends of joists or where they are supported. Sometimes one longish plank, to spread your weight, will do. Old plasterboard ceilings can 'pop' their nails when walked on a lot, especially if fixed up under older lath and plaster ceilings with nails that were too short (e.g. to repair war damage). This is not usually serious — the effect is for plaster skim to come off over the nail heads — and can be made good with filler afterwards.

Fire safety

How many households consider how they would escape from their house if it were to catch fire in the night? This is relevant to energy conservation if you make changes to windows, with double-glazing or insulating shutters.

It is unlikely that any such changes would completely prevent escape, but delay in breaking through such a window sufficiently to jump out of it could be fatal. PVC double-glazing could be in that category, unless you have thought about the material sufficiently to enable you to remove or break through it quickly and efficiently, preferably without breaking an openable window behind it, and thus causing unnecessary broken glass in the area on which you want to jump.

Glass double-glazing may be even more problematic. The best answer is an inner pane that will open inwards on hinges, with catches that are neither awkward nor stiff to handle, especially by children or in the dark. Fixed glazing is more difficult: you will need to bash at it quite a lot to make a large hole without too many sharp edges to cut people climbing through.

In short, don't put yourself in such a position that, after a tragedy, someone says 'if only he hadn't so carefully insulated those windows'. Take these factors into account when you do the insulating.

Care of existing wiring

Mention was made in the sections on roof insulation of the need to check the condition of existing wiring before insulating over the top. This is particularly important because:

● old wiring has the tendency to overheat and, with the insulation over it, it would then be more likely to catch fire;

● once you have laid the insulation, the wiring is 'out of sight, out of mind'.

Combustible materials

This issue was mentioned in discussing wall insulation, largely because such materials have already been under consideration by fire authorities in respect of public-sector housing. This may seem illogical, because in practice many people with modern upholstered furniture have in that furniture just as great a fire risk as is likely to get built into the building with insulation. But if someone dies in one of those houses and the death can in some way be attributed to a new building material in the house, then those responsible for the choice of that material will be blamed.

A further factor is that, once the house is built, the materials used in it are normally there for good. Furniture is not expected to last so long, and it can be expected that safer materials will progressively replace the present ones in due course.

A variety of plastic foams are used as insulants, offering different fire hazards. Some of them will burn fiercely and help spread and intensify a fire. Others produce dense smoke in very large quantities such that within minutes or even less it is impossible either to see or breathe, and some produce poisonous fumes such that you could die in bed without even waking up. In each case the form of prevention is the same, namely:

● shield the foam with an incombustible layer that will prevent the plastic igniting for long enough to allow people in the house to become aware of the fire and get out;

● prevent the plastic catching fire *behind* this layer, by barring a supply of air to it;

● attach the surface layer firmly enough to the wall (or ceiling) behind to prevent it collapsing inwards should the heat of the fire in front melt the plastic.

The above represent the agreed views of fire authorities. At present they have no say in what any private individual does to his own house. But it is unwise not to pay attention to their informed understanding of fire risk and how to avoid it.

Combustible insulants can also be used in roofs, but the glass and mineral fibres at present most commonly used for loft insulation are completely non-combustible. Polystyrene is not, of course, and neither is cellulose fibre in its raw state. The latter, however, ought to have been treated chemically to prevent it burning, though to date there has been some argument as to whether or not it can smoulder. There is also the problem that the chemicals used to proof it against fire are relatively expensive and in rather short supply, so that only with quite strict control of the products will the purchaser be protected from this possible (if unlikely) hazard. This said, it is doubtful whether a fire in the loftspace is anything like such a danger to the occupants of the house as one downstairs, although it could cost you the roof (or the whole building) whereas non-combustible insulation would not.

You have to face the fact that any room or house that is well insulated will suffer more from a fire, in so far as heat from the fire will build up a great deal more quickly than in a house built to conventional standards. Internal wall insulation will make the most difference in this respect.

It is just possible that in the course of time house-insurance companies will start to take an interest in DIY (or other) alterations. If they do they may take account of any increase in fire risk to the building. (In this context they will be more interested in the effect of fire on the structure than on the occupants, as it is the former they are insuring.) It would be, as far as houses are concerned, a change in their policy: at present they only want to know whether the house is of 'conventional construction', and they do not normally ask more detailed questions.

Regarding the risks and uncertainties associated with condensation behind insulating wall lining, the solution may be for insurance companies to be persuaded to insure against it. This is unlikely to occur very soon, but it certainly might if dry lining became common, and if further research were to indicate that the apparent hazards are real enough for the householder to need to buy protection. In such circumstances the fire hazards of wall insulation would probably be considered at the same time. However, these remarks are entirely speculative.

Chemical interactions

Two decades ago, the chance of any untoward chemical reaction occurring in a house was very small. But the risk has always been there whenever a new material has been introduced, especially in the hands of the DIY amateur, who is less likely than a professional designer to be able to predict in advance what will happen when one substance comes into contact with another. Warnings on product packaging can never be entirely reliable, because the range of possible reactions is immense.

The risks arising out of insulation work are not great, largely because nearly all the substances involved are dry. Possible exceptions are:

● expanded polystyrene, which can absorb the plasticiser from some sorts of PVC, so as to make it go brittle in the course of time: the main worry here is electric cable, but dusting the cable with French chalk should provide adequate protection;

● soft mastics or foams used to seal gaps: their solvents are likely to leach out slowly over a long period, so check out any possible interactions before proceeding;

● fire retardants used in cellulose-fibre insulation: these have been mentioned in the context of metal fastenings in roof structures, but beware also of possible effects on brass or copper in electric ceiling roses and in plumbing.

In all these matters it is up to you to reassure yourself beforehand and get as much information as possible. If in any doubt, rather than ask questions over the counter in a DIY shop or builder's merchant, phone the manufacturer and ask for 'technical advice'. You will then in all probability be speaking to a chemist who will know what he or she is talking about, and whose job it is to answer just such questions for the sake of the good name of the product. Get the answer in writing if you can.

Hazardous fibres

The realisation (a few years ago) of the insidious nature of blue asbestos focused attention on fibrous dust in general. The problem is that small particles of certain substances can enter the lungs and occasionally cause irritation, bronchitis and even forms of cancer.

While blue asbestos has been proved beyond doubt to be particularly dangerous, some medical research has suggested that *any* fibrous mineral should be regarded as potentially hazardous until it is proven innocent — although neither glass nor mineral fibre as used in insulation seem to have acquired a definitely bad reputation as yet, and in all probability never will. However, the precautions mentioned in Chapter 7, in talking about the use of fibrous insulation materials, are well worth observing. In particular, avoid breathing the air in the loft without a mask while handling the material; also avoid fixing it up under a roof, or using it in places where it will be rubbed or frayed later, without a dust-proof covering to prevent fall-out.

Cavity filling

Cavities were introduced into domestic wall construction some decades ago to ensure weatherproofness, and have long been compulsory in new construction in all but a few areas of Britain. They are not in fact necessary everywhere, as is demonstrated by the millions of solid-walled houses that exist, most of which are bone dry. But with some types of bricks and in some conditions (driving rain or coastal winds) water will penetrate the outer skin of a cavity wall and pour down its inside face. In these conditions anything that bridges the cavity is likely to cause damp to pass right through the wall. Building Inspectors will not permit cavity filling in such areas. Elsewhere many of them will permit it, but that does not mean that the process is proof against failure.

Many houses are quite badly built in respect of details of craftsmanship that usually do not matter, but occasionally do. The condition of wall cavities is one such detail: all too often mortar droppings are allowed to fall into the cavity during building, and build up on cavity ties or at the bottom of the cavity, occasionally to the point of forming a 'bridge' above damp-course level. These faults should never occur, but the desirable precautions are very seldom taken in practice – at least in non-exposed areas. Some badly built houses leak at once, and are put right. The majority do not leak, either because the outer skin never gets wet enough, or because the cavity is well ventilated and water crossing it dries out before it penetrates the inner skin. There are other even more serious faults that can go unnoticed because the weather does not test that particular house very severely. Such faults are built-in and invisible, and are said to be 'sub-critical', which is to say they are not evident as failures. The filling of cavities with insulation can cause sub-critical faults to become critical, not through any failure of the filling material or process, but because they interacted with dormant building faults. Probably the majority of the 'failings' of cavity filling are of this sort.

Despite the fact that the installer could not be expected to find them when inspecting the house, most people whose houses have suffered such faults have had them put right under guarantee. However, it would be very strange if installers' guarantees were to give complete protection against absolutely anything that could have been wrong with the house beforehand. It is nevertheless in installers' interests in the long term to put faults right on most occasions, otherwise people would be 'scared off' the process. A few unfortunates, however, have had problems following cavity filling and have failed to get satisfaction.

Cavity filling is akin to Russian roulette. Failures are fairly rare, but of these a minority are very expensive to remedy. The ideal form of indemnity for the householder does not seem to exist: it would be an all-risks insurance policy that would cover all filling operations carried out by firms of acceptable technical competence, given that the house were cleared as 'safe for filling' by a disinterested party (presumably the Building Inspector). Installers who demonstrated a high failure rate could have the cover withdrawn from all their future installations, and would thus go out of business.

10 Solar Energy for Houses

Perhaps houses were invented simply because people could not all have south-facing caves. The fact that the idea of a north-facing cave is so depressing demonstrates how important to domestic morale is the sun.

If in recent generations people have become less aware of sunshine as a useful heater of houses, it is no doubt because most of our houses are built to standardised designs, and placed on the ground in relation to the street and the garden rather than the sun. There was an era, too, when people tried to shut out the sun because it tended to bleach fabrics and wallpaper, and in any case coal for heating was cheap and plentiful.

Since the demise of 'back-to-back' housing nearly all houses have received sufficient sunlight to avoid depression, although it has not been regarded as a means of heating. For some years, too, it has been normal, in the design of flats, to observe rules of 'sunlighting' for reasons of amenity. It is not customary for heating systems to be designed to exploit the incidental heat of the sun, and controls that will do this automatically are relatively new.

The houses we live in make a lot of difference to our attitudes. We are conscious of the sun in houses where it makes its presence felt sufficiently to affect our behaviour. Only a fortunate few have a living-room with a large south-facing window which the sun alone will warm in mid-winter. With push-button warmth available in every room the sun's heat hardly seems very important, and very few architects try, or have the chance, to design houses in which sunshine is exploited to its full potential as a form of heating. But times are changing in that respect: a select few among recent houses have been designed for the 'passive' exploitation of solar heat.

Practical possibilities

However little your house allows for the exploitation of sunshine, there are nevertheless things you can do to improve its practically useful effects. Some of these were described in Chapter 5, but this chapter goes somewhat further.

While the use of sunshine falling on a building's structure or admitted through windows is usually described as 'passive' solar collection, the incorporation of special collectors and some means of transferring the heat to where it is more useful is described as 'active' collection. The former is usually cheaper to incorporate in a design for a new building; the latter is easier to reconcile with other aspects of the design, and indeed easier to add to an existing building. A variety of techniques have been tried, mostly with some success. Naturally such devices are not so effective in Britain as in sunnier places, but Britain does have nearly two-thirds as many hours of sunshine daily in summer as in the United States and Australia, though only a quarter as many in the winter.

By far the commonest 'active' technique is the 'flat-plate' collector, which is basically a static arrangement oriented and tilted in a direction that is best, in most cases, for round-the-year performance. The collector receives the sun at the intensity at which it shines, i.e. with no attempt to intensify its rays. The receiving surface is treated to absorb as much as possible of the heat, protected by glazing on top to reduce re-radiation, and insulation underneath to prevent further waste. The transfer medium (normally water or air) passes under, over or through the receiving surface, and thence to wherever the heat is to be used or stored.

Sun-heated water

The most promising uses for flat-plate collectors in Britain so
far have been for swimming pool heating — already quite common — and for domestic hot water supply. The latter use, in a
well-designed installation, tends to achieve 50—60 per cent
reduction in annual hot water costs, most of this being en-
joyed after February and before November.

Whether such a saving is worthwhile depends greatly on how
much the installation costs, and on the price of the fuel being
saved. It is generally reckoned that a fairly good DIY installation
can in fact be cost-effective where the fuel saved is on-peak
electricity. Solar heating *might* be cost-effective as a substitute
for a cheaper fuel if the household uses a great deal of hot
water. If you want to go in for such an installation make sure
you get good, experienced advice on all aspects of its design, as
there are a host of quite serious mistakes you might otherwise
make. Also make quite sure that the solar apparatus is compatible with your present system for heating domestic hot water.

As important as the details of the collecting apparatus and its
controls are those of its fixing to the building. Remember, too,
that it will need to be kept clean, as dirt reduces its performance.
You may need to get the consent of your local authority in
respect of its effect on the appearance of the house, and of the
materials used (e.g. combustible plastic glazing); i.e. both
planning *and* Building Regulations consent may be necessary.

There are now a number of firms in Britain marketing solar-
collector installations. There have been incompetent and even
unscrupulous firms operating in the past, but more recently the
Solar Trade Association has become established as a means of
improving competence and protecting the customer. But
collector installations are quite expensive, and you should
investigate your present hot water costs so as to understand
whether you would be doing it to save money or in truth largely
for fun. Although fuel costs will almost certainly continue to
rise faster than general inflation, remember that the solar
installation may have maintenance costs, and probably a limited
life, too.

Space heating by flat-plate collectors

For space heating, 'active' solar collection has rather less chance of success, but some ideas have already been tested and found promising. Development is less advanced than with domestic hot water installations, and success is to be measured more in terms of whether the installation works than on its economic prospects: the latter will depend on the cost of mass-produced installations of a proven design. The main problem is that of storing heat between sunny spells, or even between seasons. Water is the cheapest storage medium, though dry materials may be more convenient. Water can absorb more heat in relation to its bulk than any other cheap and chemically inactive substance. To last out between sunny days you need about five cubic metres of water for heat storage, and to last between seasons about 40 cubic metres (about 1400 cubic feet).

The heat-transfer medium can be air or water, each with its own advantages. Air is more difficult to employ in an existing building, unless it has certain useful characteristics by chance. Some schemes make use of heat pumps (see Chapter 11) to 'upgrade' the heat available, but these can be quite complicated and are at their least effective in cold weather. Few solar space-heating systems can make do without some supplementary form of heat for long periods of cold weather or overcast skies. Mains gas is an especially suitable fuel for this purpose if the boiler can be used to inject heat into an exhausted heat store, in which case the boiler can be at its most efficient. Off-peak electricity would also be suitable, although at present slightly more expensive.

If solar space heating or even domestic water heating were to become very popular, with mains gas or electricity as the back-up fuel, there would of course be a saving in energy overall, but the prices of these fuels would tend to rise because very awkward peaks in demand would be produced. It is always cheaper to produce piped or wired energy when the demand is relatively steady. For this reason, perhaps the ultimate solution for solar space heating would be systems incorporating large storage capacity that could be 'topped up' in adverse conditions by a solid-fuel stove, perhaps using domestic refuse, or by home-generated methane gas. Both these fuels are cost-free, but they can be produced only in small quantities by the average household. (It is possible that these variants have already been tried; there is no complete register of such projects.)

Irrespective of the amount of money available to develop solar technology, it takes time for teams of experts to be assembled; then each project takes maybe three or four years to design, build, monitor and tune up to its peak performance. Hence the pace of development must inevitably be slow.

Heat pumps and concentrating collectors

As an example of the problems to be overcome in developing the technology of solar collection, consider the fact that the hotter the surface of a substance is, the less heat it will absorb, because the rate of heat transfer depends on temperature difference. This law of physics applies throughout solar systems wherever heat has to be collected, or transferred from one substance to another. It favours only the circumstances of supplying heat to a nearly exhausted heat store, but as the store accumulates heat, the circulating water or air will get hotter and the activities of collection and 'accumulation' will slow down. It follows that a large store that never gets very hot is best. Hence for space heating, a heat-distribution medium at low temperature will get most from a relatively cool store.

Because of such problems, many of the later experiments with 'active' solar energy for houses have also incorporated heat pumps, which enable heat from a large quantity of relatively cool material to be concentrated in a smaller quantity of hotter material. However, heat pumps add to the system's running cost, noise and maintenance liability.

To some extent heat-transfer problems are relieved if the temperature of the water at the point of collection is higher. This can be achieved by focusing the sun's rays by flat or curved reflectors. Large focusing reflectors were first used to generate steam for mechanical power in America as long ago as 1878.

To work properly, however, focusing reflectors need to be adjusted as the sun's angle changes through the day and through the seasons, so they are of little use for domestic purposes unless inconveniently complicated.

Simple DIY possibilities

Chapter 5 included a number of tips on household management which amount to advice on the exploitation of passive solar energy without modifying the house at all. However, many houses offer opportunities for 'passive' or 'semi-passive' applications whereby heat gains from sunshine can be considerably increased.

Sun-rooms and conservatories

Anyone who has a greenhouse, a conservatory, or in fact any sun-room, will know how much heat it can collect. (In fact recent research has shown that a conservatory can save more energy than a solar collector, for the same cost.) It also protects that wall of the house from outside air and reduces heat loss. For this reason a lean-to glazed addition is of more use than a flat-roofed sun-room, in that it covers a larger area of wall. Such rooms have a double advantage, in sunny weather during the heating season: they fill up with hot air, which warms the wall of the house behind them, and they allow the sun to warm that wall directly far more effectively than it would in the open air. Chapter 3 explains the reason for this: the glass of the glazed lean-to will prevent the warmed wall from re-radiating heat back to the sky. This principle has been described for many years as 'greenhouse effect'.

The same effect could equally be exploited by simply glazing over an outside solid brick wall so that sunshine warms it more effectively. It sounds worth trying — preferably starting with a smallish panel (say one metre square) first. Fix it firmly so that wind does not rip it off. If the method works, this wall will get warmer on the inside on cold but sunny days, but you must allow time for the warmth to pass through the wall. Check the need for local-authority consents.

As a practical example of a glazed lean-to, *Figure 10.1* shows such an outbuilding on the author's own house. It covers an unheated storage/work area; the roof faces east, and catches the sun from dawn until midday. On a sunny day in early March the outside air temperature was 5°C, and inside it was 12°C at chest height and 20°C near the apex of the glass lean-to. Heavy brick walls store the heat for the rest of the day. In midwinter the enclosure protects the side of the house from east winds, and is always 2–5°C above outside air temperature without sunshine.

Fig. 10.1 Glazed lean-to, heated by 'passive' collection of solar heat

A glazed outbuilding could be exploited more effectively if the warm air in it could be admitted to the house (*Figure 10.2*). An airbrick near the top of the lean-to would tend to let the warm air through into the house under the force of its own buoyancy, unless wind pressures acting throughout the house were to dictate otherwise. Such warm air would be of most

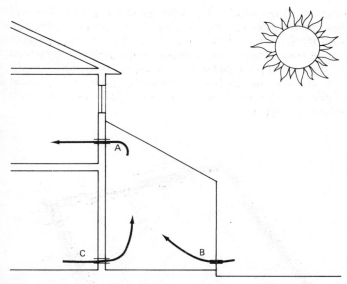

Fig. 10.2 Using conservatory to heat house: warmest air collects near apex (A), and can be either introduced to first floor by buoyancy or fed to another part of house by fan and duct. Conservatory should be well sealed except for low-level inlet (B), or alternatively a recirculating vent from house (C)

use if it were warmer than that in the house, but would still be of use if it could be made to replace other ventilation inputs that would otherwise consist of cold outside air. Whether or not this is worth trying to achieve depends on what opportunities the house happens to offer. A fan may well be necessary, and some means of controlling the rate of air and hence its

temperature. It will be necessary to close off the flow completely in warm weather, and also to draughtseal the outbuilding
well, with the exception of a single small opening near the
ground to feed the flow of air up into the house.

The roofspace

Another opportunity offers itself on the occasion of a roof
needing to be re-tiled or slated. If instead of re-covering it with
another slate or tile finish you were to glaze over all or part of
a south-facing slope, the loft-space would collect a great deal
of solar heat that would otherwise just be radiated back to the
sky. However this is done it is almost bound to cost a great
deal more than the cheapest tiles, and a little more than blue
asbestos-cement slates.

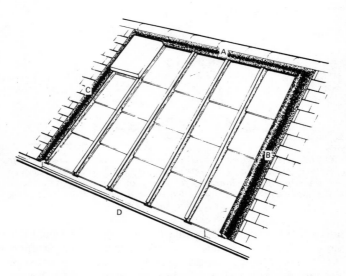

*Fig. 10.3 Roof surface replaced by patent glazing, for active or passive
solar collection. All edge details ABCD need careful design to be
permanently weatherproof: if in doubt get advice*

A 'patent glazing' system (see Appendix 2) offers the best means of glazing. Leave the bare rafters in position after stripping tiles and battens, then fit the patent-glazing bars. Ordinary glass is acceptable over narrow spans between bars. Use short sheets and lap them, inserting a putty bedding under the edges where lapping them leaves most of the edge unsupported. All this can be done from inside the loft, if you form a trapdoor in place of the last sheet. This will allow you to poke your head and shoulders through on occasions later to wash down the glass with a long mop. See *Figure 10.3*.

Such a glazed roof offers several possibilities, including:
• the installation of flat plate collectors underneath; or
• constructing an air-medium 'active' collector, by forming voids under the glass (between joists) with a sheet of black polythene stapled to the underside of the joists (*Figure 10.4*) and insulated underneath with fibreboard.

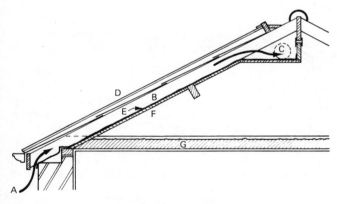

Fig. 10.4 Air-medium solar collector (variation on Fig. 10.3 scheme):
air enters at eaves (A), passes up through collector (B) to ridge
channel (C), thence by fan and duct to house. Collector consists of
patent glazing (D), voids (B) between rafters, and black polythene
collector surface (E) backed by fibre-board insulation (F). Continue
polythene down to eaves in case of winter condensation on glass;
don't forget roof insulation (G)

For the former, detailed advice would be necessary on the design of a proper system, as explained earlier. The latter is simpler, and the design of the air-handling system should not present problems. The warm air could be pumped down into the house with a fan. One of the kind built for internal bathrooms would serve excellently, and some of these are designed to use ordinary plastic 100 mm (4 in) diameter drainpipe as ducting. This would have to be passed down into the house to a convenient outlet.

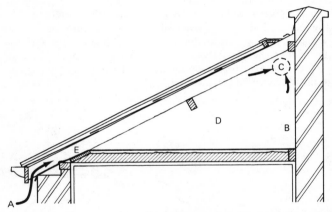

Fig. 10.5 Variation on Fig. 10.3, using whole space of monopitch loft and heat-storage capacity of solid party wall (B): cool air enters at A and is extracted into house at (C). Vital to block off ends of space (D): polythene (E) used to throw off condensation drips only. Ordinary slate roof, instead of glazing, gives much reduced performance (for less cost), although collecting efficiency of slates would be improved by laying corrugated clear plastic over them: this must be very well fixed down, and top edge must stop some way short of party wall for Building Regulations reasons

A monopitch roof with a party wall at the high side offers another variant, in which the heat-storage capacity of the wall can be used. In this case, use building paper or polythene walls to section off the part of the roof forming the collector, and collect from any position high up on this wall. See *Figure 10.5*.

Cover any mineral-fibre insulation in this section with poly-thene, to prevent its dust being pumped down into the house. Preferably use an east or south-east facing roof so as to store morning sun.

If you don't want to glaze the roof, it may be practicable to achieve approximately the same effect under a tiled or slated roof (slates, being black, will get hotter in the sun). Try the idea out by sheeting over just one joist space first, and test the temperature of air passing up (by buoyancy) through the tunnel so formed. Do this regularly on sunny days through early spring until the end of the heating season and you will find out whether or not you have a viable form of supplementary space heating available for parts of the heating season. You will need to ensure that your main heating system can cut itself back at such times.

The glazed version of the collector is sure to work, because as much heat will be available as would go into a flat-plate collector in the same location. However, data from a collector design book will enable you to calculate roughly whether enough heat is available to justify the cost of the glazing, fan and ducting. All these arrangements would be controlled by a thermostat adapted to cut in when the air was warm enough to use and cut out when it was not. The viability of the alternative air-under-the-slates proposal will probably depend on there not being too much air leakage between slates. It may be possible to seal some of these with a gun mastic. In any case, the exploratory experiment will be easy, cheap and fun.

You should be warned that any of these arrangements, but particularly that with glass, will almost certainly lead to surface condensation, from the underside of the glass or slates, collecting on the polythene sheet below and running down it. A small gutter and drainpipe may be necessary to get rid of this water.

11 The Future

Solving problems is one of mankind's favourite pastimes, and if starved of problems we tend to invent them. However, there is nothing invented about the prospect of the world running out of cheap energy, so thousands of people the world over have been applying their ingenuity to it, with a single-minded devotion that is completely international. This is not just a phenomenon of the industrial nations: there is no hope of the world supporting its present population, distributed as it is, without harnessing natural energy to help us.

There is a broad range of problems to solve. We must try to extend existing energy supplies (i.e. go on looking for more coal, oil and natural gas), solve the remaining problems of newer technology (such as how to make nuclear power efficient, cheap and altogether safe), explore newer possibilities (the whole range of solar sciences, geothermal energy, wind and wave power) and, last but not least, continue to pursue ways of using energy that allow us to do more with less. Conservation is indeed one of our most important energy resources.

This chapter covers very briefly an enormous range of subjects that bear indirectly on domestic energy conservation. It touches on those inventions most likely to become familiar in domestic life over the next ten or twenty years, and on developments in science and technology that seem at present most likely to affect energy supply in the medium and long term.

New house design and construction

A host of 'low energy' houses have been built over the last few years, most of them in some part successful. Together they have helped to clarify which ideas are most practical, cheapest to build, easiest to adapt to, and most acceptable to live with. For example, a 'zero energy' house that cost virtually nothing to run, but would cost twice as much as an ordinary house to build, would be of little more than academic interest. To come even more down to earth, it would be equally impracticable to build houses that, in order to work properly and to last long enough, would take standards of craftsmanship not normally available. It would take at least twenty years for the house-building industry to adapt. It would be 'possible' in less time of course, but the need would have to be very desperate for the whole community to agree to making the necessary resources available. In democratic societies there is a large gulf between what is possible and what is practicable.

Nevertheless we *shall* change our ways of building new houses. They will be a lot better insulated, they will have to have very good control of ventilation and at the same time be a lot more draughtproof, and (because slightly different forms of construction will be needed) special care and science will be needed in their detailed design. The latter two problems are very great, because thousands of designers, contractors, foremen and builders will have to learn new tricks, albeit only modifications of their present ways of working. If this change is not brought about slowly and with care, we may be storing up building failures for ten years hence at the rate of a quarter of a million a year. People spend 20–30 years paying for their houses, and expect them to last 50–100 years: this is something on which our whole 'system' depends.

Some of these new houses will have solar hot water, and perhaps a few will have solar space heating. Probably far more normal will be houses designed to use 'passive' solar gain to reduce heating costs. Heating systems will be different, too: they will be much smaller, better controlled, and many more of them fired by solid fuel, of which we seem to have far larger reserves than of other fuels. They will be of new design, of course, requiring the minimum of attention, and clean and convenient. Such developments are already well advanced. It is likely that lighting and household appliances will be far more efficient, too, and we may learn to rely on cold-water detergents to wash clothes and dishes. Domestic wastes and refuse will probably be ordered differently, with emphasis on the recycling of useful materials, such as paper, plastics, metals and glass, and the conversion of sewage and degradable refuse into useful energy, either by efficient burning or by digestion into methane gas and fertiliser, to a far greater extent than is done already.

Heat pumps

There is little doubt that the heat pump has a future in domestic life. It is a machine very similar to a refrigerator. If you loosen up your understanding of the refrigerator you will be able to grasp that it takes heat out of the food put into the cabinet, and releases it from the coils at the back, in a far more concentrated form. What the heat pump does is to re-frigerate air or water or earth (or waste water) and introduce the heat from them, in a more concentrated form, into the house for space heating and hot water.

There is always heat available from the environment to put into a house, although a certain amount of energy is needed to achieve the process of concentrating it. Most heat pumps are, like refrigerators, electric. A good one can produce 3–4 kW of useful space heating for the expenditure of 1 kW of electricity. The extra heat comes usually from the environment, but don't worry about heat pumps making the weather colder. Remember all the heat you put into the house eventually leaks out again. This does affect the weather, in that large towns tend to be a degree or two warmer than the surrounding countryside: if all buildings were heated by heat pumps this difference would lessen a little.

Heat pumps tend to struggle somewhat in the coldest weather, when more heat is wanted but less is available, so most houses designed for them will require a little boost from elsewhere as well. This will create peaks in demand for other fuels − a problem that is not insoluble but will have to be dealt with when it arrives.

Heat reclaim

Another possibility that is exercising a lot of minds is to 're-claim' as much as possible of the heat lost by a building, from those places where the loss is most concentrated. The obvious targets are the warm air lost in ventilation, and all the hot water that goes down the drain. Use of the first is hampered at present by the difficulty of making houses airtight enough to allow the outgoing warm air to be concentrated, rather than lost through a thousand tiny cracks and gaps in construction.

The heat from waste hot water is of more use, though it would be easier to get at if we simply kept all waste water in a large tank and pulled the heat out with a heat pump. But this takes energy, and the equipment is quite expensive (even if mass-produced), needs maintenance, and wears out or goes wrong.

Either sort of heat reclaim might be cheaper to achieve without such mechanical complication if we just relied on simple heat exchange. However, as you may remember from Chapter 3, the best you will get will be warm air or water at a temperature half-way between that of the fresh cold and the waste warm. Hence at the absolute maximum of efficiency this procedure could only reclaim half the heat wasted, and in practice probably a good deal less.

There may be scope for DIY experimentation along the following lines, for example: run the wastes from sink, basins and bath into a tank or a length of pipe in contact with the coldest surface that usefully can be warmed. Choose a surface that will cause the warmth to be stored for long enough to be useful. Remember that most waste water is dirty, so any such system may tend to clog up in time, as the suspended scum settles out.

Aerogenerators (windmills)

There is a lot of power in the wind, and it has been used by man for centuries. Modern knowledge of aerodynamics has improved our methods of using it. The problems, however, have been:

● Efficient equipment is relatively expensive, though fairly homespun versions can be made a lot more cheaply.

● Because wind is intermittent, you need to store the energy unless you can use it straight away (for example to pump water— not a common domestic task). Storage usually means batteries, even the cheapest of which are costly, bulky and vulnerable to the stresses of this kind of intermittent loading.

● Strong winds have the habit of tearing down windmills: experience has proved that they can be really quite dangerous.

● Windmills suffer from the turbulence of air around buildings and that created by trees.

● They are noisy and cause vibration; so, apart from the turbulence problem, they need to be sited a good distance away from buildings.

Large-scale aerogenerators are, however, being investigated in Holland and elsewhere, for public electricity supply. In theory they have possibilities to offer, as long as people can accept their noise and their conspicuous appearance. Many of the most modern types have a vertical axis, so there is no gearing at the top of a tower to waste energy, and nothing basic has to move when the wind changes direction.

For reference, a two-blade (horizontal axis, propeller-type) aerogenerator two metres (six feet) in diameter will produce about 40 watts of electricity at a wind speed of 15 km/h (10 miles/h), 600 watts or more at 40 km/h (25 miles/h). The former is the *average* windspeed in the UK, but it is better at a favourable site of course. If you need 3000 kWh a year, you need to generate continuously around the year at 350 watts. Converting the d.c. electricity stored in batteries into the 250 V a.c. needed by most appliances tends to be inefficient.

One has to conclude that aerogenerators are perhaps a viable proposition for remote rural locations only. The energy could be stored as heat, but what we are really looking for is a cheap source of electricity, which is so much more adaptable, and so expensive at present.

Energy from waste

Methane gas can be produced from animal waste (sewage) with an admixture of vegetable waste to eke it out. This process is used a great deal in India, where there are between two and three thousand 'Gobar Gas' plants for making gas from cow-dung and other animal wastes (India has a research institute to further the adoption of these plants). In the UK, sewage is used in several large sewage-treatment plants to make methane to run the diesel-powered generators needed to serve the sewage works' own needs.

So the technology for making methane from sewage already exists. But the waste produced by an average household is pathetically inadequate to its energy requirements, which makes the idea of a home methane plant entirely impracticable unless there is waste from a number of other animals to supple-ment it. On a farm, therefore, the picture may be entirely different. Note that, after extracting the methane, the residue is even better as a fertiliser than it was before. The gas produced is of rather lower calorific value than mains natural gas, but can still be used for cooking, heating, lighting and running diesel engines.

For those interested in this technology, the most reliable source of practical advice is the Indian Gobar Gas Research Institute.

Solar energy — a range of sciences

The use of solar heat on a large scale involves physics and engineering, and at its most sophisticated includes power-station-size solar concentrating-collector installations such as exist in the USA and Europe. These are at development stage but seem likely to be commercially viable for such countries as the USA and France — the latter having limited resources of indigenous energy but a fair amount of both land and sunshine in the south. The scope for exploitation of solar heat is quite considerable, as the technology of lenses and mirrors is being brought to bear on the problem of focusing sunshine in simpler and cheaper ways than having a large, heavy reflector that tracks the sun through the day. Also, the heat, however collected, can be harnessed to a variety of pumps and engines and thus turned into mechanical or electrical energy.

The USA has been the leading country in the development of 'solar cells'. These follow the principle used in photographic light meters, in which light energy is converted directly into electricity. There are no moving parts, the cells last a long time, and they respond equally to diffused light (e.g. from an overcast sky) simply in proportion to the light available. The problem is that the cells are expensive to make at present, although frequently using fairly cheap materials (various types have been tried). The basic technology involves semiconductors — the materials from which transistors and 'micro-chips' are made — but despite enormous efforts and great expenditure the problems of mass-producing solar cells cheaply have yet to be solved. The present cells can be made to yield more power by concentrating the sun's rays on them, but then the problem of tracking arises all over again.

The dream, as yet unrealisable, is that one day houses will be able to have their roof surfaces covered with solar cells and thus generate all the electricity they need during daylight hours. Some storage will of course be needed as well — and here the problem of batteries (preferably superior to the present lead/acid car batteries) recurs, as with aerogenerators. Here again, years of research have so far failed to solve the problem. The answer may well arrive, if it ever does, by accident, perhaps by a discovery in another field of pure science, rather than by dogged pursuit of a particular Holy Grail.

Agricultural and biological possibilities

Fossil fuels provide energy when burned because they are concentrates of substances created from organic material. We can burn wood, straw and leaves, too, but generally with less reward in proportion to the density of the material. But it is not just this that matters; the biological nature of the potential fuel is relevant — for example, some grasses are better fuel than others. What we need now are forms of plant life that reproduce very quickly and take up solar energy in a concentrated form that can readily be used by man. Unfortunately plants that grow fastest tend to grow in water and produce very wet material. This cannot be efficiently burned because of the energy needed to drive off the water. Also, some plant life that appears on the face of it to yield a lot of usable energy requires a great deal of expensive fertiliser to grow densely in the UK soil and climate.

Deeper thought has led to the possibility of exploring the chemistry of plant life and using plants as a source of high-energy fuels, such as methane, alcohol or certain oils (available from some nuts, sugar and certain trees). Another possibility is to use otherwise inaccessible (i.e. difficult to convert into energy) materials to feed other plantlife that is more useful, such as certain forms of algae. It is believed that potatoes, from which it is quite easy to make alcohol, could be grown in greenhouses at about four times their present yield, and thus as an energy crop yield a profit upon the cost of greenhouses (perhaps heated with power-station waste heat). This sort of consideration is important in assessing the potential of energy crops, because any proposal to 'grow' energy in the UK would compete for the use of land with our already inadequate food production.

Methane is produced by bacterial action on 'waste' matter, alcohol by a chemical process upon some constituents of the plant (i.e. the conversion of carbohydrate into hydrocarbon). A yet more 'straightforward' possibility involves certain algae, which under particular conditions, and using solar energy, will produce hydrogen gas from water. This surely is the most exciting possibility of all. The process is not easy to achieve using naturally occurring organisms, but it is thought possible to synthesise organic compounds that will achieve the same effect. If we could achieve this we should be working upon the very basic components of living matter such as those that originally led, over thousands of millions of years, to the development of life on earth.

Wave-power

Indirect products of the sun's energy are the wind, waves and tide. We have discussed wind-power, and tidal possibilities are already quite well known since the one major possibility in the UK — the Severn estuary — has been thoroughly discussed in the Press.

Wave-power has received serious attention over the past few years only, but is of particular relevance to British geography and climate. To harness it we should need large floating devices — vessels or articulated chains of rafts — anchored temporarily or permanently in Atlantic waters off the west coasts. The theory seems to be workable, and to date water-tank tests have performed well enought to justify scaled-down prototypes to be built for testing in inshore waters.

The principal problems seem to be once again those of storage, and the favoured method appears to be the manufacture of hydrogen gas on the vessel for transport ashore. There will be dangers to shipping from the generating devices, of course, but probably less serious than those presented by supertankers. As to the transport of hydrogen gas, this does give cause for anxiety, unless the gas can safely be brought inshore and converted into electricity more or less immediately.

Continuing importance of conservation

With all these possibilities, in addition to that offered by nuclear power, it seems unlikely that we shall lack sufficient energy in the long-term future. But all this research and development has been undertaken because we are certain that prices will rise sufficiently to justify new sources of energy — and these, though perhaps inexhaustible and non-polluting (with some exceptions), will nevertheless cost quite a lot more than the fossil fuels we are used to. So it will certainly pay to spend money and effort and ingenuity on reducing our consumption. This may or may not lead to a lower standard of living or lower economic growth. It will not necessarily lead to a lower quality of life — in fact the reverse may be the case.

Appendix 1
Further Reading

General Government publications:
BRE Working Party Report CP 56/75, HMSO, 1975
Advisory Council on Energy Conservation, Paper 7, HMSO, 1978
Energy Policy – a Consultative Document (Cmnd 7101), HMSO, 1978 (the Energy Green Paper)
A Warmer House at Lower Cost, edited by Dr P. J. V. Agius, Watt Committee on Energy (Ltd), 1979
Compare Your Home Heating Costs (booklet free from Department of Energy)

Lighting design:
IES code for interior lighting, Illuminating Engineering Society (London), 1977
Interior lighting design (4th ed), Lighting Industry Federation Ltd and Electricity Council, 1973

Solar energy:
Solar energy – a UK assessment, International Solar Energy Society, 1976
The solar house, P. R. Sabady, Newnes–Butterworths, 1978
Department of Energy, *Solar energy: its potential contribution within the United Kingdom,* Energy Paper 16, HMSO, 1976
Solar heating for domestic hot water (a guide to good practice), Heating and Ventilating Contractors Association, 1978
Code of practice for solar water heating, Solar Trade Association, 1978

Natural energy in general:

How to use natural energy, The Natural Energy Association (Kingston-on-Thames)

Handbook of home-made power, Bantam Books, 1974

Condensation:

The control of condensation in dwellings (BS 5250), British Standards Institution, 1975

Domestic Energy Note 4 — Condensation, Working Party on Heating and Energy, DOE Housing Development Directorate, 1979

For advice on:

All building problems: Building Research Establishment, Garston, Watford, Herts

Gas heating and other appliances: your Regional Gas Board

Solid fuel: Solid Fuel Advisory Service (there are eight Regional Centres)

Electric heating and other appliances: your local Electricity Board.

Appendix 2
Glossary of Terms

Balanced flue: (relatively new) flue arrangement for gas appliance; exhausts fumes and draws in combustion air through a single terminal — usually on the external wall immediately behind the appliance.

Ball valve: water valve as in w.c. cistern, cold water tank or central-heating feed tank; opens by dropping of a floating ball whenever level drops below 'full' level, and closes when that level is recovered.

Breather paper: a building paper that keeps out rain but lets moisture vapour escape.

Cellulose fibre: soft, light, fluffy insulating material made from waste newspaper.

Circulator: small gas heater for hot water only, to heat water in a storage cylinder. (A circulator is usually smaller than a central heating boiler.)

Conduction: transfer of heat through a substance, without any movement of the substance (as through the handle of a metal spoon whose bowl is in a hot liquid).

Convection: transfer of heat via a gas or liquid (e.g. air or water) due to currents in it caused by the effect of hotter gas or liquid 'floating' on cooler gas or liquid.

Diverter valve: control on central-heating/hot-water system to direct most of the boiler's heat to one purpose or the other, or in between, according to choice.

Dew point: the precise temperature at which moisture in humid air condenses. This varies according to the 'vapour pressure' of the air, which can be measured by relating its relative humidity to its temperature, read at the same time.

Draughtstripping: foam, rubber, plastic or metal strip to exclude draughts around doors, windows, hatches etc.

Exfoliated vermiculite: granular insulating material made by heating naturally occurring layered rock, causing it to puff up into lightweight granules.

Expansion pipe: pipe from a central heating or hot water system allowing for expansion (or boiling) of water in the system to be relieved by pouring into a high-level tank. Hot water/CH (radiator) systems usually have two: one for domestic hot water, the other for radiators and cylinder heating coil.

Fluorescent light: electric light in white glowing tube form, common in shops, offices etc., as well as in some domestic kitchens.

Gigajoule: unit of energy equivalent to approximately 278 kilowatt-hours ('units') of electricity or 9.5 therms of gas.

Gravity feed: supply of heat to hot water cylinder (common) or central heating system (now rare) by convection currents, through large pipes. There is no pump.

Heat pump: appliance like a refrigerator in reverse, to take low-grade heat from one place and turn it into high-grade heat in another (e.g. outside air to heat a house).

Hygrometer: simple instrument to indicate relative humidity of air, using the tendency of hair or paper to expand in humid air.

Interstitial condensation: condensation that forms *inside* a wall, roof or floor, as vapour permeates from a warm side towards a cooler side and falls to dew-point temperature somewhere within the thickness.

Kilowatt-hour: unit of energy or electricity equivalent to the consumption of one kilowatt for one hour (usually written kW h).

Lumen: unit of light output, useful for comparing the brightness of light sources (bulbs or tubes). Ordinary light bulbs produce around 12 lumens per watt of electricity used.

Mastic: soft, sticky material used for sealing cracks against rain or wind (in outside conditions); usually applied from a tube or 'mastic gun'.

Mineral fibre: light, fluffy material made from glass or other mineral, as used commonly for insulation.

Monopitch roof: sloping roof consisting of *one* flat slope, rather than the commoner kind with two slopes meeting at a ridge. A 'lean-to' is a monopitch with its top edge against a wall going up higher.

Multi-point water heater: an 'instantaneous' water heater serving several taps at once, and heating the water supply to any (or all) taps as soon as they are turned on, but not storing any hot water.

Munsell value: a figure used to describe the amount of light a particular colour (usually of paint) reflects. See also *Reflectance factor*, and page 96.

Negative radiation: the effect whereby a cold surface (or void, such as outer space) absorbs heat radiated by an object and hence makes it cold. Especially applied to the coldness of large windows or walls, which can cause discomfort.

Patent glazing: a system of roofing (or walling) in glass using a manufactured section of metal (usually aluminium) which supports the glass and has some means of rapidly weatherproofing over its edges. These patent glazing 'bars' are laid down the slope, usually parallel to each other, with glass fitted between them.

Plasterboard: manufactured rigid board of plaster between sheets of tough paper. Commonly used to line partitions, ceilings and sometimes external walls of houses. Often finished with a 'skim coat' of plaster. (It is the commonest material for 'dry lining'.) Can be obtained with foam plastic insulation bonded to the back face.

Plasticiser: non-evaporating chemical in a soft plastic material (such as PVC used for cable sheathing). In some circumstances it can 'migrate' from one plastic to another, very slowly, leaving the former one brittle.

Polystyrene (expanded): white rigid foam plastic, cheapest and commonest for insulation.

Pumped circuit: circular system of piping containing water driven round by a pump, to distribute heat (common in radiator central-heating systems).

'Pumped primary': hot water system with the heat given to the cylinder by a pumped circuit, rather than by gravity, which is more common.

Pyrolitic: describes an oven with a cleaning facility achieved by heating to a high enough temperature to burn off greasy dirt.

Radiation: transfer of heat by invisible waves, through 'transparent' materials (including air and vacuum), with little or no heat being absorbed by the 'transparent' material. All objects with temperatures above absolute zero $(-273°C)$ radiate heat, and thus transfer it to others of a lower temperature. Transparency of materials varies according to source temperature of heat being transmitted; e.g. glass is transparent to heat from the sun but not to heat from within a room. Hence 'greenhouse effect'.

Reflectance factor: an expression of the capacity of a shade of colour to reflect (rather than absorb) light. It is the percentage of white light that is reflected. See also *Munsell Value*.

Sarking: sheet material (usually felt or polythene, but may be timber or insulation board) laid under tiles or slates to help keep out the weather, and sometimes to stiffen the roof structure.

Silicone: transparent treatment used to make porous materials (such as brick) water-repellent. It does not change the appearance of the material it is brushed on to.

Size: water-soluble glue usually used to make porous surfaces (e.g. plaster or wallpaper) absorb paint less readily. Commonly in form of crystals, like brown sugar, to be dissolved in hot water.

Solar collector: any device used to collect heat radiated by the sun and transfer it to a convenient medium for use inside a building (or engine) for use to heat water supply, or space, or to drive machinery. As applied to buildings, 'active' collection describes the above principle using air or water as a transfer medium; 'passive' collection refers to the use of direct sunlight for space heating, possibly stored in the building structure; 'flat-plate' collectors involve a collecting surface that receives sunshine at its natural intensity; 'concentrating' collectors involve lenses or mirrors, to achieve higher temperatures over a smaller area.

Suspended floor: the 'hollow' timber ground floor commonest in older houses in England (but still normal in Scotland), with a ventilated air space underneath.

Temperature gradient: air temperature difference within a room, most commonly between floor and ceiling as created by convection currents.

Thermal mass: the capacity of an object to absorb, and thus store, heat. (High thermal mass = large capacity; low thermal mass = small capacity.)

Thermal transmittance: capacity of a material to conduct heat.

Thermostatic control: a control of heating or hot water whose task is to maintain a predetermined temperature (see text).

Trickle ventilation: principle of providing a small flow of air for ventilation (far smaller than that achieved by the minimum opening of windows); and closable in windy conditions to offset incurable structural air leakage.

Trussed rafters: timber roof construction in which each rafter or pair of rafters is part of a prefabricated supporting member (more common in recent construction).

Tungsten light: lighting with tungsten filament – the common electric light bulb used in houses.

Urea-formaldehyde: the commonest and cheapest plastic foam used for insulating walls by cavity filling.

'U-value': the thermal transmittance of parts of a building, expressed in heat units per unit of area and degree of temperature difference between one side and the other.

Voidless roof: roof with no loft space (e.g. with ceiling up against the underside of the rafters).

Watt: small unit of power, or of electricity; e.g. a 100-watt light bulb uses 100 watts of power at the prescribed voltage.

Appendix 3
Author's Own Improvements

Figure A.1 shows how the author's gas and electricity consumption has changed since 1972, and the improvements that achieved the savings are listed below. Gas consumption has been reduced by 19 per cent, electricity by 63 per cent (there has been some shift from electricity to gas). Although the total annual cost increased from £193 in 1972 to £283 in 1978, the latter amount would have been £472 without the improvements. This shows a saving of 40 per cent over the seven years — and there are still improvements to be made.

Roof: 100 mm (4 in) glass fibre generally, but storage area floored (no insulation) and enclosed in a 'tent' of foil-backed building paper stapled to joists. Tanks within storage area. (Roof slated, with no sarking.)

Walls: warmest rooms dry-lined with battens (rot-proofed), 25 mm (1 in) cavity, foil-backed building paper and plasterboard. (Walls 225 mm (9 in) brick with soft plaster.)

Windows: largest windows double-glazed in glass, all fixed glazing with extra glass beaded in from the inside over airtight glazing seal. Large bays triple-glazed in glass-clear PVC in removable timber frames fixed with double-glazing clips (removable in summer). Louvre ventilators untreated.

Floors: most ground floor areas have new floor finish over existing softwood board floors, effectively curing draught gaps between boards and at skirtings.

	1972				1973				1974				1975				1976				1977				1978			
Gas (therms) Quarter totals	414	273	50	327	390	227	32	319	425	194	53	289	350	200	80	302	392	157	72	258	341	217	71	243	383	178	70	239
Year total	1064				968				961				932				880				872				870			
% of 1972	100				91				90				88				83				82				82			
Electricity (kW h) Quarter totals	2549	2324	1801	2846	2586	2863	2382	1672	3678	2403	2094	2431	3295	2076	554	984	1300	631	501	1115	801	606	1012	1417	969	753	866	
Year total	9520				9503				10586				6909				3547				3747†				4005†			
% of 1972	100				99				111				72				37				39				42			
Gigajoules total*	107 (100%)				100 (93%)				104 (97%)				89 (83%)				73 (68%)				73 (68%)				74 (69%)			

Annotations (by year):

- 1973: 50 mm roof insulation; Progressive wall insulation to 3 rooms
- 1975: Draught-stripping; ▲ Boiler change (summer hot water changed to gas)
- 1976: ▲ First double glazing; ▲ Another 50 mm roof insulation; ▲ 'Solar' glazed extension added (incl. 1 extra heated room)
- 1977: ▲ More draught-stripping
- 1978: ▲ Some double-glazing made triple

Prices in 1972 (6.825 p/therm, 0.945 p/unit):		1972 consumption at 1978 prices: (22.8/15.3 p/therm, 2.9 p/unit):		1978 consumption at 1978 prices:	
Gas	£ 72.62	Gas	£178.38	Gas	£148.71
+ st. ch.	23.08	+ st. ch.	8.00	+ st. ch.	8.00
Elect.	89.96	Elect.	276.08	Elect.	116.15
+ st. ch.	7.40	+ st. ch.	10.52	+ st. ch.	10.52
	£193.06		£472.98		£283.38

* Gigajoules are a universal unit of energy (Gas figures adjusted for assumed efficiency on all gas appliances of 65 per cent).
1 Gigajoule = 9.5 therms or 278 kW h.
† Extra electric lighting used in 1977–1978.

Fig. A.1 Author's fuel consumption and costs, 1972–1978

Heating/hot water: original 45 000 BTU boiler replaced by 38 000 BTU balanced-flue low-thermal-capacity boiler, whereupon summer hot water provided by gas rather than electricity. Push-button programmer makes control easy. Immersion heater used on occasions for rapid top-up.

Passive solar gain: glazed lean-to over 'back yard' provides permanently warm zone at 10 m length of outside wall up to first-floor window sill level, and some daytime heating to most of house on sunny spring and autumn days (see *Figure 10.1*). Large south-facing windows to rear living-room provide daytime solar heat to this room.

Appendix 4
Sources of Materials

The following materials are new or relatively difficult to obtain. These particular products are not necessarily the cheapest or best of their kind.

Bitumen-impregnated fibre board: information available from FIDOR (Fibre Building Board Development Organisation Ltd), 6 Buckingham Street, London WC2N 6BZ (01-839 1122).

Draughtstrip: two new products, ribbed rubber strip (from Sweden) and nylon pile type (from Spain), both from College Housewares, Bristol.

Expanded polystyrene slab: information about uses and supply from Vencel Resil Ltd, West Street, Erith, Kent DA8 1DD (Erith 36922). Note that expanded polystyrene is normally the cheapest foam plastic insulant: it is not the safest, nor does it have the best insulating value per unit of thickness, but it is probably best value for money in most applications.

Foil-backed building papers: all information from British Sisalkraft Ltd, (St Regis Coating & Laminating Division), Knight Road, Strood, Rochester, Kent ME2 2AW (Medway 74171).

Gap-filling aerosol foam: 'FEB Handy Foam' by FEB (Great Britain) Ltd, Albany House, Swinton Hall Road, Swinton, Manchester M27 1DT (061-794 7411). Try large builders' merchants first.

Glass-clear PVC: Transatlantic Plastics Ltd (EM459), Garden Estate, Ventnor, Isle of Wight (Ventnor 852241), by mail order or from ten retail outlets in the South of England.

Insitu sealed-unit double-glazing: the 'Signa' system of Affa Produkter A/S, Broenge 4, DK 2635 1SHØJ, Denmark. This firm has expressed the intention of establishing distributors in Britain.

Insulating blinds: 'Thermoblind' Insulated Window Shutters Ltd, 28 Queen Street, Huddersfield HD1 2SP (Huddersfield 27686). 13 mm blinds at 20 mm from glass give window U-value of 0.9 W/m² °C.

Plaster-surfaced polystyrene slabs: 'Perm-U-Board' by Permo-decs (Liverpool), 35 Boundary Drive, Hunts Cross, Liverpool L25 0QD (051-486 0971).

Roof ventilators: Ubbink (UK) Ltd, 108 Churchill Road, Bicester, Oxfordshire OX6 7XD (Bicester 41482).

Smokeless-fuel burning woodstove: the 'Chappée', available from Colin Brownlow & Co, 75 High Street, Egham, Surrey TW20 9HE (Egham 3595).

'Trickle' ventilators: a variety of products available from Titon Hardware Ltd, 91 London Road, Copford, Colchester, Essex CO6 1LG (Colchester 211411).

Wall insulation by 'sandwich' board: 'Gyproc Thermal Board', information from British Gypsum Ltd, Ferguson House, 15/17 Marylebone Road, London NW1 5JE (01-486 1282).

Wax-type anti-rust spray: 'Waxoyl' from Finnegan's Ltd (EMW), Eltringham Works, Prudhoe, Northumberland (Prudhoe 32411).

Appendix 5 Typical U-Values

Element before	U-value before (W/m² °C)	Treatment	U-value after (W/m² °C)
Walls			
225 mm (9 in) solid brick, 18 mm (³/₄ in) plaster	1.9	1. 20 mm battens (1 in planed), lined with foil and 9 mm (³/₈ in) plasterboard	1.1
If constantly wet would be:	2.2	2. Plasterboard over 25 mm (1 in) expanded polystyrene (or equivalent in manufactured 'sandwich' board)	0.8
		3. Damp wall after silicone treatment externally only	1.9 (?)
		4. External cladding on 25 mm (1 in) battens (forming cavity)	1.4 (with foil: 1.1)
		5. As 4, but including 25 mm (1 in) expanded polystyrene board	0.8
		6. As 4, but including 50 mm (2 in) expanded polystyrene board	0.5
275 mm (11 in) cavity walls:			
(a) brick/cavity/brick, 12 mm (¹/₂ in) plaster	1.5	1. Cavity fill with urea-formaldehyde	0.5
		2. Cavity fill with mineral fibre (safer in exposed areas)	0.5

Element before	U-value before (W/m² °C)	Treatment	U-value after (W/m² °C)
(b) brick/cavity/insulating block, 12 mm (¹/₂ in) plaster	1.0	1. As 1 above 2. As 2 above	0.4 0.4
Roofs			

(a) Unsarked roof with slates	up to 3.0	1. Glass or mineral fibre in loft: 60 mm
		                                                    *80 mm
		                                                    100 mm
		
		* or equivalent – see list on page 103
			0.6 (a or b)
			0.4 (a or b)
			0.3 (a or b)

(b) Sarked (felted) roof with tiles	about 2.0	2. 75 mm (3 in) polystyrene slabs between rafters	0.4 (a or b)
		3. Foil-backed building paper on rafters	1.5 (a) 1.2 (b)
Flat or voidless roof: 20 mm boarded deck, 100 mm joists, 9 mm plasterboard and skim	1.6	1. 6 mm (¹/₄ in) expanded polystyrene tiles	1.2
		2. 2mm expanded polystyrene surface insulation	1.5
		3. 25 mm (1 in) expanded polystyrene and plaster skim coat	0.75
		4. 25 mm expanded polystyrene and 9 mm (³/₈ in) plasterboard, or equivalent in manufactured 'sandwich' board	0.7

Windows

Single-glazed (moderate exposure)	4.3
1. Double-glazing with cavity no larger than 20 mm (little difference down to 12 mm)	2.8
2. Triple-glazing	2.0
3. Proprietary insulated shutters 13 mm thick (high-performance insulant)	0.9
4. Insulated shutters 25 mm expanded polystyrene	0.9

Floors

(a) Suspended floor: 20 mm (1 in) boards, no cover	0.8
(b) Solid floor: thin PVC tiles on concrete base	1.0
Both (a) and (b): 12.7 mm ($\frac{1}{2}$ in) fibre insulating board with 6 mm ($\frac{1}{4}$ in) total hardboard finish (insulating board to be bitumen-impregnated *or* covered with polythene vapour-check membrane).	0.7(a) 0.8(b)

Index